Praise for *Quantum Universe*

"An engaging whistle-stop tour that takes us from the birth of the field to present-day tests of the Standard Model . . . in lucid and occasionally droll prose."—*Wall Street Journal*

"In Britain, Brian Cox . . . has become the front man for physics . . . Cox has secured a large fan base with . . . his happy turns of phrase and his knack for presenting complex ideas using simple analogies. He also admirably shies away from dumbing down . . . The authors' love for their subject shines through the book."—*The Economist*

"A comprehensive account of how quantum mechanics works and why it must be real . . . The book offers many rewards, and both the language and content are so carefully chosen that it feels concise."—*New Scientist*

"A solid introduction to the 'inescapable strangeness' of the subatomic world."—*Nature*

"Thanks to his television presentations of science, Brian Cox has become Britain's best-known physics professor. No one communicates the wonders of the universe on screen with more charm, enthusiasm, and accessibility. Cox's latest, co-authored with Jeff Forshaw, a fellow physics professor at Manchester University, retains the charm and enthusiasm . . . A serious, thorough book about quantum theory for the general reader."—*Financial Times*

"Understanding the quantum world in the deep and satisfying way that we'd like to is not at all necessary to describe its workings with exquisite precision. It is this astonishingly accurate mathematical description of the consequences of a set of rules that, as strange as they may seem, actually work, that is the focus of Professors Cox and Forshaw's brief excursion . . . *The Quantum Universe* may not demystify quantum theory, but it does give the reader an idea of the size of the mountain the book is trying to climb—and a toe-hold or two to help get us started on our own ascent."—*New York Journal of Books*

"Beginning with a brief scientific history that will be familiar to anyone who's completed college physics (but accessible to those who have not), Cox and Forshaw . . . go on to explain the origin of the periodic table, strong and weak nuclear forces, 'Why We Don't Fall Through the Floor,' and myriad other interesting topics."—*Publishers Weekly*

"There is no shortage of popular introductions, but curious readers will not regret choosing this meticulous account . . . Space cannot be empty; matter constantly appears and vanishes. If quantum laws do not forbid something from happening, it will eventually happen. These are facts; experiments confirm them. Writers often explain these in relentlessly nontechnical language that converts them into a magic show, but Cox and Forshaw will have none of this. Using ingenious pedagogical examples, they demonstrate that weird quantum phenomena make perfect sense . . . An ambitious explanation of the vast quantum universe aimed at readers willing to work."—*Kirkus Reviews*

BRIAN COX & JEFF FORSHAW
THE QUANTUM UNIVERSE
(AND WHY ANYTHING THAT CAN HAPPEN, DOES)

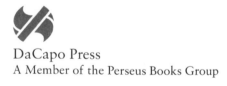

DaCapo Press
A Member of the Perseus Books Group

Typeset by Jouve (UK), Milton Keynes
Set in Dante 12/14.75pt

Cataloging-in-Publication data for this book is available from the Library of Congress.
First Da Capo Press edition 2012
First Da Capo Press paperback edition 2013
Reprinted by arrangement with Allen Lane, an imprint of Penguin Books
ISBN 978-0-306-81964-3 (hardcover)
ISBN 978-0-306-82144-8 (paperback)
ISBN 978-0-306-82060-1 (e-book)
Library of Congress Control Number 2011942393

Published by Da Capo Press
A Member of the Perseus Books Group
www.dacapopress.com

Da Capo Press books are available at special discounts for bulk purchases in the U.S. by corporations, institutions, and other organizations. For more information, please contact the Special Markets Department at the Perseus Books Group, 2300 Chestnut Street, Suite 200, Philadelphia, PA 19103, or call (800) 810-4145, ext. 5000, or e-mail special.markets@ perseusbooks.com.

10 9 8 7 6 5 4 3 2 1

Contents

Acknowledgements

We'd like to thank the many colleagues and friends who helped us 'get things right' and provided a great deal of valuable input and advice. Particular thanks go to Mike Birse, Gordon Connell, Mrinal Dasgupta, David Deutsch, Nick Evans, Scott Kay, Fred Loebinger, Dave McNamara, Peter Millington, Peter Mitchell, Douglas Ross, Mike Seymour, Frank Swallow and Niels Walet.

We owe a great debt of gratitude to our families – to Naomi and Isabel, and to Gia, Mo and George – for their support and encouragement, and for coping so well in the face of our preoccupations.

Finally, we thank our publisher and agents (Sue Rider and Diane Banks) for their patience, encouragement and very capable support. A special thanks is due to our editor, Will Goodlad.

1. Something Strange Is Afoot

Quantum. The word is at once evocative, bewildering and fascinating. Depending on your point of view, it is either a testament to the profound success of science or a symbol of the limited scope of human intuition as we struggle with the inescapable strangeness of the subatomic domain. To a physicist, quantum mechanics is one of the three great pillars supporting our understanding of the natural world, the others being Einstein's theories of Special and General Relativity. Einstein's theories deal with the nature of space and time and the force of gravity. Quantum mechanics deals with everything else, and one can argue that it doesn't matter a jot whether it is evocative, bewildering or fascinating; it's simply a physical theory that describes the way things behave. Measured by this pragmatic yardstick, it is quite dazzling in its precision and explanatory power. There is a test of quantum electrodynamics, the oldest and most well understood of the modern quantum theories, which involves measuring the way an electron behaves in the vicinity of a magnet. Theoretical physicists worked hard for years using pens, paper and computers to predict what the experiments should find. Experimenters built and operated delicate experiments to tease out the finer details of Nature. Both camps independently returned precision results, comparable in their accuracy to measuring the distance between Manchester and New York to within a few centimetres. Remarkably, the number returned by the experimenters agreed exquisitely with that computed by the theorists; measurement and calculation were in perfect agreement.

This is impressive, but it is also esoteric, and if mapping the miniature were the only concern of quantum theory, you might be forgiven for wondering what all the fuss is about. Science, of course, has no brief to be useful, but many of the technological and social changes that have revolutionized our lives have arisen out of fundamental research carried out by modern-day explorers whose only motivation is to better understand the world around them. These curiosity-led voyages of discovery across all scientific disciplines have delivered increased life expectancy, intercontinental air travel, modern telecommunications, freedom from the drudgery of subsistence farming and a sweeping, inspiring and humbling vision of our place within an infinite sea of stars. But these are all in a sense spin-offs. We explore because we are curious, not because we wish to develop grand views of reality or better widgets.

Quantum theory is perhaps the prime example of the infinitely esoteric becoming the profoundly useful. Esoteric, because it describes a world in which a particle really can be in several places at once and moves from one place to another by exploring the entire Universe simultaneously. Useful, because understanding the behaviour of the smallest building blocks of the Universe underpins our understanding of everything else. This claim borders on the hubristic, because the world is filled with diverse and complex phenomena. Notwithstanding this complexity, we have discovered that everything is constructed out of a handful of tiny particles that move around according to the rules of quantum theory. The rules are so simple that they can be summarized on the back of an envelope. And the fact that we do not need a whole library of books to explain the essential nature of things is one of the greatest mysteries of all.

It appears that the more we understand about the elemental nature of the world, the simpler it looks. We will, in due course, explain what these basic rules are and how the tiny building blocks conspire to form the world. But, lest we get too dazzled by the underlying simplicity of the Universe, a word of caution is in order: although the basic rules of the game are simple, their consequences are not necessarily easy to calculate. Our everyday experience of the world is dominated by the relationships between vast collections of many trillions of atoms, and to try to derive the behaviour of plants and people from first principles would be folly. Admitting this does not diminish the point – all phenomena really are underpinned by the quantum physics of tiny particles.

Consider the world around you. You are holding a book made of paper, the crushed pulp of a tree.[1] Trees are machines able to take a supply of atoms and molecules, break them down and rearrange them into cooperating colonies composed of many trillions of individual parts. They do this using a molecule known as chlorophyll, composed of over a hundred carbon, hydrogen and oxygen atoms twisted into an intricate shape with a few magnesium and nitrogen atoms bolted on. This assembly of particles is able to capture the light that has travelled the 93 million miles from our star, a nuclear furnace the volume of a million earths, and transfer that energy into the heart of cells, where it is used to build molecules from carbon dioxide and water, giving out life-enriching oxygen as it does so. It's these molecular chains that form the superstructure of trees and all living things, and the paper in your book. You can read the book and understand the words because you have eyes that can convert the scattered light from the pages into electrical impulses that are interpreted by your brain, the most complex structure we know of in the Universe. We have discovered that all these things are nothing more than assemblies of atoms, and that the wide variety of atoms are constructed using only three particles: electrons, protons and neutrons. We have also discovered that the protons and neutrons are themselves made up of smaller entities called quarks, and that is where things stop, as far as we can tell today. Underpinning all of this is quantum theory.

The picture of the Universe we inhabit, as revealed by modern physics, is therefore one of underlying simplicity; elegant phenomena dance away out of sight and the diversity of the macroscopic world emerges. This is perhaps the crowning achievement of modern science; the reduction of the tremendous complexity in the world, human beings included, to a description of the behaviour of just

---

1. Unless of course you are reading an electronic version of the book, in which case you will need to exercise your imagination.

a handful of tiny subatomic particles and the four forces that act between them. The best descriptions we have of three of the forces, the strong and weak nuclear forces that operate deep within the atomic nucleus and the electromagnetic force that glues atoms and molecules together, are provided by quantum theory. Only gravity, the weakest but perhaps most familiar of the four, does not at present have a satisfactory quantum description.

Quantum theory does, admittedly, have something of a reputation for weirdness, and there have been reams of drivel penned in its name. Cats can be both alive and dead; particles can be in two places at once; Heisenberg says everything is uncertain. These things are all true, but the conclusion so often drawn – that since something strange is afoot in the microworld, we are steeped in mystery – is most definitely not. Extrasensory perception, mystical healing, vibrating bracelets to protect us from radiation and who-knows-what-else are regularly smuggled into the pantheon of the possible under the cover of the word 'quantum'. This is nonsense born from a lack of clarity of thought, wishful thinking, genuine or mischievous misunderstanding, or some unfortunate combination of all of the above. Quantum theory describes the world with precision, using mathematical laws as concrete as anything proposed by Newton or Galileo. That's why we can compute the magnetic response of an electron with such exquisite accuracy. Quantum theory provides a description of Nature that, as we shall discover, has immense predictive and explanatory power, spanning a vast range of phenomena from silicon chips to stars.

Our goal in writing this book is to demystify quantum theory; a theoretical framework that has proved famously confusing, even to its early practitioners. Our approach will be to adopt a modern perspective, with the benefit of a century of hindsight and theoretical developments. To set the scene, however, we would like to begin our journey at the turn of the twentieth century, and survey some of the problems that led physicists to take such a radical departure from what had gone before.

Quantum theory was precipitated, as is often the case in science, by the discovery of natural phenomena that could not be explained by the scientific paradigms of the time. For quantum theory these were many and varied. A cascade of inexplicable results caused excitement and confusion, and catalysed a period of experimental and theoretical innovation that truly deserves to be accorded that most clichéd label: a golden age. The names of the protagonists are etched into the consciousness of every student of physics and dominate undergraduate lecture courses even today: Rutherford, Bohr, Planck, Einstein, Pauli, Heisenberg, Schrödinger, Dirac. There will probably never again be a time in history where so many names become associated with scientific greatness in the pursuit of a single goal; a new theory of the atoms and forces that make up the physical world. In 1924, looking back on the early decades of quantum theory, Ernest Rutherford, the New-Zealand-born physicist who discovered the atomic nucleus in Manchester, wrote: 'The year 1896 . . . marked the beginning of what has been aptly termed the heroic age of Physical Science. Never before in the history of physics has there been witnessed such a period of intense activity when discoveries of fundamental importance have followed one another with such bewildering rapidity.'

But before we travel to nineteenth-century Paris and the birth of quantum theory, what of the word 'quantum' itself? The term entered physics in 1900, through the work of Max Planck. Planck was concerned with finding a theoretical description of the radiation emitted by hot objects – the so-called 'black body radiation' – apparently because he was commissioned to do so by an electric lighting company: the doors to the Universe have occasionally been opened by the prosaic. We will discuss Planck's great insight in more detail later in the book but, for the purposes of this brief introduction, suffice to say he found that he could only explain the properties of black body radiation if he assumed that light is emitted in little packets of energy, which he called 'quanta'. The word itself means 'packets' or 'discrete'. Initially, he thought that this was purely a mathematical trick, but subsequent work in 1905 by Albert Einstein on a phenomenon called the photoelectric effect gave

5

further support to the quantum hypothesis. These results were suggestive, because little packets of energy might be taken to be synonymous with particles.

The idea that light consists of a stream of little bullets had a long and illustrious history dating back to the birth of modern physics and Isaac Newton. But Scottish physicist James Clerk Maxwell appeared to have comprehensively banished any lingering doubts in 1864 in a series of papers that Albert Einstein later described as 'the most profound and the most fruitful that physics has experienced since the time of Newton'. Maxwell showed that light is an electromagnetic wave, surging through space, so the idea of light as a wave had an immaculate and, it seemed, unimpeachable pedigree. Yet, in a series of experiments from 1923 to 1925 conducted at Washington University in Saint Louis, Arthur Compton and his co-workers succeeded in bouncing the quanta of light off electrons. Both behaved rather like billiard balls, providing clear evidence that Planck's theoretical conjecture had a firm grounding in the real world. In 1926, the light quanta were christened 'photons'. The evidence was incontrovertible – light behaves both as a wave and as a particle. That signalled the end for classical physics, and the end of the beginning for quantum theory.

2.   Being in Two Places at Once

Ernest Rutherford cited 1896 as the beginning of the quantum revolution because this was the year Henri Becquerel, working in his laboratory in Paris, discovered radioactivity. Becquerel was attempting to use uranium compounds to generate X-rays, discovered just a few months previously by Wilhelm Röntgen in Würzburg. Instead, he found that uranium compounds emit 'les rayons uraniques', which were able to darken photographic plates even when they were wrapped in thick paper that no light could penetrate. The importance of Becquerel's rays was recognized in a review article by the great scientist Henri Poincaré as early as 1897, in which he wrote presciently of the discovery 'one can think today that it will open for us an access to a new world which no one suspected'. The puzzling thing about radioactive decay, which proved to be a hint of things to come, was that nothing seemed to trigger the emission of the rays; they just popped out of substances spontaneously and unpredictably.

In 1900, Rutherford noted the problem: 'all atoms formed at the same time should last for a definite interval. This, however, is contrary to the observed law of transformation, in which the atoms have a life embracing all values from zero to infinity.' This randomness in the behaviour of the microworld came as a shock because, until this point, science was resolutely deterministic. If, at some instant in time, you knew everything it is possible to know about something, then it was believed you could predict with absolute certainty what would happen to it in the future. The breakdown of this kind of predictability is a key feature of quantum theory: it deals with probabilities rather than certainties, not because we lack absolute knowledge, but because some aspects of Nature are, at their very heart, governed by the laws of chance. And so we now understand that it is simply impossible to predict when a particular atom will decay. Radioactive decay was science's first encounter with Nature's dice, and it confused many physicists for a long time.

Clearly, there was something interesting going on inside atoms, although their internal structure was entirely unknown. The key discovery was made by Rutherford in 1911, using a radioactive source to bombard a very thin sheet of gold with a type of radiation known as alpha particles (we now know them to be the nuclei of helium atoms). Rutherford, with his co-workers Hans Geiger and Ernest Marsden, discovered to their immense surprise that around 1 in 8,000 alpha particles did not fly through the gold as expected, but bounced straight back. Rutherford later described the moment in characteristically colourful language: 'It was quite the most incredible event that has ever happened to me in my life. It was almost as incredible as if you fired a 15-inch shell at a piece of tissue paper and it came back and hit you.' By all accounts, Rutherford was an engaging and no-nonsense individual: he once described a self-important official as being 'like a Euclidean point: he has position without magnitude'.

Rutherford calculated that his experimental results could be explained only if the atom consists of a very small nucleus at the centre with electrons orbiting around it. At the time, he probably had in mind a situation similar to the planets orbiting around the Sun. The nucleus contains almost all the mass of the atom, which is why it is capable of stopping his '15-inch shell' alpha particles and bouncing them back. Hydrogen, the simplest element, has a nucleus consisting of a single proton with a radius of around 1.75×10^{-15} m. If you are unfamiliar with this notation, this means 0.00000000000000175 metres, or in words, just under two thousand million millionths of a metre. As far as we can tell today, the single electron is like Rutherford's self-important official, point-like, and it orbits around the hydrogen nucleus at a radius around 100,000 times larger than the nuclear diameter. The nucleus has a positive electric charge and the electron has a negative electric charge, which means there is an attractive force between them analogous to the force of gravity that holds the Earth in orbit around the Sun. This in turn means that atoms are largely empty space. If you imagine a nucleus scaled up to the size of a tennis ball, then the tiny electron would be smaller than a mote of dust orbiting at a distance of a kilometre. These figures are quite surprising because solid matter certainly does not feel very empty.

Rutherford's nuclear atom raised a host of problems for the physicists of the day. It was well known, for instance, that the electron should lose energy as it moves in orbit around the atomic nucleus, because all electrically charged things radiate energy away if they move in curved paths. This is the idea behind the operation of the radio transmitter, inside which electrons are made to jiggle and, as a result, electromagnetic radio waves issue forth. Heinrich Hertz invented the radio transmitter in 1887, and by the time Rutherford discovered the atomic nucleus there was a commercial radio station sending messages across the Atlantic from Ireland to Canada. So there was clearly nothing wrong with the theory of orbiting charges and the emission of radio waves, and that meant confusion for those trying to explain how electrons can stay in orbit around nuclei.

A similarly inexplicable phenomenon was the mystery of the light emitted by atoms when they are heated. As far back as 1853, the Swedish scientist Anders Jonas Ångstrom discharged a spark through a tube of hydrogen gas and analysed the emitted light. One might assume that a glowing gas would produce all the colours of the rainbow; after all, what is the Sun but a glowing ball of gas? Instead, Ångstrom observed that hydrogen emits light of three very distinct colours: red, blue-green and violet, like a rainbow with three pure, narrow arcs. It was soon discovered that each of the chemical elements behaves in this way, emitting a unique barcode of colours. By the time Rutherford's nuclear atom came along, a scientist named Heinrich Gustav Johannes Kayser had compiled a six-volume, 5,000-page reference work entitled *Handbuch der Spectroscopie*, documenting all the shining coloured lines from the known elements. The question, of course, was why? Not only 'why, Professor Kayser?' (he must have been great fun at dinner parties), but also 'why the profusion of coloured lines?' For over sixty years the science of

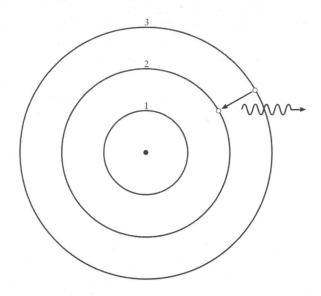

Figure 2.1: Bohr's model of an atom, illustrating the emission of a photon (wavy line) as an electron drops down from one orbit to another (indicated by the arrow).

spectroscopy, as it was known, had been simultaneously an observational triumph and a theoretical wasteland.

In March 1912, fascinated by the problem of atomic structure, Danish physicist Niels Bohr travelled to Manchester to meet with Rutherford. He later remarked that trying to decode the inner workings of the atom from the spectroscopic data had been akin to deriving the foundations of biology from the coloured wing of a butterfly. Rutherford's solar system atom provided the clue Bohr needed, and by 1913 he had published the first quantum theory of atomic structure. The theory certainly had its problems, but it did contain several key insights that triggered the development of modern quantum theory. Bohr concluded that electrons can only take up certain orbits around the nucleus with the lowest-energy orbit lying closest in. He also said that electrons are able to jump between these orbits. They jump out to a higher orbit when they receive energy (from a spark in a tube for example) and, in time, they will fall back down, emitting light in the process. The colour of the light is determined directly by the energy difference between the two orbits. Figure 2.1 illustrates the basic idea; the arrow represents an electron as it jumps from the third energy level down to the second energy level, emitting light (represented by the wavy line) as it does so. In Bohr's model, the electron is only allowed to orbit the proton in one of these special, 'quantized', orbits; spiralling inwards is simply forbidden. In this way Bohr's model allowed him to compute the wavelengths (i.e. colours) of light observed by Ångstrom – they were to be attributed to an electron hopping from the fifth orbit down to the second orbit (the violet light), from the fourth orbit down to the second (the blue-green light) or from the third orbit down to the second (the red light). Bohr's model also correctly predicted that there should be light emitted as a result of electrons hopping down to the first orbit. This light is in the ultra-violet part of the spectrum, which is not visible to the human eye, and so it was not seen by Ångstrom. It had, however, been spotted in 1906 by Harvard physicist Theodore Lyman, and Bohr's model described Lyman's data beautifully.

Although Bohr did not manage to extend his model beyond hydrogen, the ideas he introduced could be applied to other atoms. In particular, if one supposes that the atoms of each element have a unique set of orbits then they will only ever emit light of certain colours. The colours emitted by an atom therefore act like a fingerprint, and astronomers were certainly not slow to exploit the uniqueness of the spectral lines emitted by atoms as a way to determine the chemical composition of the stars.

Bohr's model was a good start, but it was clearly unsatisfactory: just why were electrons forbidden from spiralling inwards when it was known that they should lose energy by emitting electromagnetic waves – an idea so firmly rooted in reality with the advent of radio? And why are the electron orbits quantized in the first place?

And what about the heavier elements beyond hydrogen: how was one to go about understanding their structure?

Half-baked though Bohr's theory may have been, it was a crucial step, and an example of how scientists often make progress. There is no point at all in getting completely stuck in the face of perplexing and often quite baffling evidence. In such cases, scientists often make an ansatz, an educated guess if you like, and then proceed to compute the consequences of the guess. If the guess works, in the sense that the subsequent theory agrees with experiment, then you can go back with some confidence to try to understand your initial guess in more detail. Bohr's ansatz remained successful but unexplained for thirteen years.

We will revisit the history of these early quantum ideas as the book unfolds, but for now we leave a mass of strange results and half-answered questions, because this is what the early founders of quantum theory were faced with. In summary, following Planck, Einstein introduced the idea that light is made up of particles, but Maxwell had shown that light also behaves like waves. Rutherford and Bohr led the way in understanding the structure of atoms, but the way that electrons behave inside atoms was not in accord with any known theory. And the diverse phenomena collectively known as radioactivity, in which atoms spontaneously split apart for no discernible reason, remained a mystery, not least because it introduced a disturbingly random element into physics. There was no doubt about it: something strange was afoot in the subatomic world.

The first step towards a consistent, unified answer is widely credited to the German physicist Werner Heisenberg, and what he did represented nothing less than a completely new approach to the theory of matter and forces. In July of 1925, Heisenberg published a paper throwing out the old hotchpotch of ideas and half-theories, including Bohr's model of the atom, and ushered in an entirely new approach to physics. He began: 'In this paper it will be attempted to secure the foundations for a quantum theoretical mechanics which is exclusively based on relations between quantities which in principle are observable.' This is an important step, because Heisenberg is saying that the underlying mathematics of quantum theory need not correspond to anything with which we are familiar. The job of quantum theory should be to predict directly observable things, such as the colour of the light emitted from hydrogen atoms. It should not be expected to provide some kind of satisfying mental picture for the internal workings of the atom, because this is not necessary and it may not even be possible. In one fell swoop, Heisenberg removed the conceit that the workings of Nature should necessarily accord with common sense. This is not to say that a theory of the subatomic world shouldn't be expected to accord with our everyday experience when it comes to describing the motion of large objects, like tennis balls and aircraft. But we should be prepared to abandon the prejudice that small things behave like smaller versions of big things, if this is what our experimental observations dictate.

There is no doubt that quantum theory is tricky, and absolutely no doubt that Heisenberg's approach is extremely tricky indeed. Nobel Laureate Steven Weinberg, one of the greatest living physicists, wrote of Heisenberg's 1925 paper:

> If the reader is mystified at what Heisenberg was doing, he or she is not alone. I have tried several times to read the paper that Heisenberg wrote on returning from Heligoland, and, although I think I understand quantum mechanics, I have never understood Heisenberg's motivations for the mathematical steps in his paper. Theoretical physicists in their most successful work tend to play one of two roles: they are either sages or magicians ... It is usually not difficult to understand the papers of sage-physicists, but the papers of magician-physicists are often incomprehensible. In that sense, Heisenberg's 1925 paper was pure magic.

Heisenberg's philosophy, though, is not pure magic. It is simple and it lies at the heart of our approach in this book: the job of a theory of Nature is to make predictions for quantities that can be compared to experimental results. We are not mandated to produce a theory that bears any relation to the way we perceive the world at large. Fortunately, although we are adopting Heisenberg's philosophy, we shall be following Richard Feynman's more transparent approach to the quantum world.

We've used the word 'theory' liberally in the last few pages and, before we continue to build quantum theory, it will be useful to take a look at a simpler theory in more detail. A good scientific theory specifies a set of rules that determine what can and cannot happen to some portion of the world. They must allow predictions to be made that can be tested by observation. If the predictions are shown to be false, the theory is wrong and must be replaced. If the predictions are in accord with observation, the theory survives. No theory is 'true' in the sense that it must always be possible to falsify it. As the biologist Thomas Huxley wrote: 'Science is organized common sense where many a beautiful theory was killed by an ugly fact.' Any theory that is not amenable to falsification is not a scientific theory – indeed one might go as far as to say that it has no reliable information content at all. The reliance on falsification is why scientific theories are different from matters of opinion. This scientific meaning of the word 'theory', by the way, is different from its ordinary usage, where it often suggests a degree of speculation. Scientific theories may be speculative if they have not yet been confronted with the evidence, but an established theory is something that is supported by a large body of evidence. Scientists strive to develop theories that encompass as wide a range of phenomena as possible, and physicists in particular tend to get very excited about the prospect of describing everything that can happen in the material world in terms of a small number of rules.

One example of a good theory that has a wide range of applicability is Isaac Newton's theory of gravity, published on 5 July 1687 in his *Philosophiæ Naturalis Principia Mathematica*. It was the first modern scientific theory, and although it has subsequently been shown to be inaccurate in some circumstances, it was so good that it is still used today. Einstein developed a more precise theory of gravity, General Relativity, in 1915.

Newton's description of gravity can be captured in a single mathematical equation:

$$F = G\frac{m_1 m_2}{r^2}$$

This may look simple or complicated, depending on your mathematical background. We do occasionally make use of mathematics as this book unfolds. For those readers who find the maths difficult, our advice is to skip over the equations without worrying too much. We will always try to emphasize the key ideas in a way that does not rely on the maths. The maths is included mainly because it allows us to really explain why things are the way they are. Without it, we should have to resort to the physicist-guru mentality whereby we pluck profundities out of thin air, and neither author would be comfortable with guru status.

Now let us return to Newton's equation. Imagine there is an apple hanging precariously from a branch. The consideration of the force of gravity triggered by a particularly ripe apple bouncing off his head one summer's afternoon was, according to folklore, Newton's route to his theory. Newton said that the apple is subject to the force of gravity, which pulls it towards the ground, and that force is represented in the equation by the symbol F. So, first of all, the equation allows you to calculate the force on the apple if you know what the symbols on the right-hand side of the equals sign mean. The symbol r stands for the distance between the centre of the apple and the centre of the Earth. It's r^2 because Newton discovered that the force depends on the square of the distance between the objects. In non-mathematical language, this means that if you double the distance between the apple and the centre of the Earth, the gravitational force drops by a factor of 4. If you triple the distance, it drops by a factor of 9. And so on. Physicists call this behaviour an inverse square law. The symbols m_1 and m_2 stand for the mass of the apple and the mass of the Earth, and their appearance encodes Newton's recognition that the gravitational force of attraction between two objects depends on the product of their masses. That then begs the question: what is mass? This is an interesting question in itself, and for the deepest answer available today we'll need to wait until we talk about a quantum particle known as the Higgs boson. Roughly speaking, mass is a measure of the amount of 'stuff' in something; the Earth is more massive than the apple. This kind of statement isn't really good enough, though. Fortunately Newton also provided a way of measuring the mass of an object independently of his law of gravitation, and it is encapsulated in the second of his three laws of motion, the ones so beloved of every high school student of physics:

1. Every object remains in a state of rest or uniform motion in a straight line unless it is acted upon by a force;
2. An object of mass m undergoes an acceleration a when acted upon by a force F. In the form of an equation, this reads $F = ma$;
3. To every action there is an equal and opposite reaction.

Newton's three laws provide a framework for describing the motion of things under the influence of a force. The first law describes what happens to an object when no forces act: the object either just sits still or moves in a straight line at constant speed. We shall be looking for an equivalent statement for quantum particles later on, and it's not giving the game away too much to say that quantum particles do not just sit still – they leap around all over the place even when no forces are present. In fact, the very notion of 'force' is absent in the quantum theory, and so Newton's second law is bound for the wastepaper basket too. We do mean that, by the way – Newton's laws are heading for the bin because they have been exposed as only approximately correct. They work well in many instances but fail totally when it comes to describing quantum phenomena. The laws of quantum theory replace Newton's laws and furnish a more accurate description of the world. Newton's physics emerges out of the quantum description, and it is important to realize that the situation is not 'Newton for big things and quantum for small': it is quantum all the way.

Although we aren't really going to be very interested in Newton's third law here, it does deserve a comment or two for the enthusiast. The third law says that forces come in pairs; if I stand up then my feet press into the Earth and the Earth responds by pushing back. This implies that for a 'closed' system the net force acting on it is zero, and this in turn means that the total momentum of the system is conserved. We shall use the concept of momentum throughout this book and, for a single particle, it is defined to be the product of the particle's mass and its speed, which we write $p = mv$. Interestingly, momentum conservation does have some meaning in quantum theory, even though the idea of force does not.

For now though, it is Newton's second law that interests us. $F = ma$ says that if you apply a known force to something and measure its acceleration then the ratio of the force to the acceleration is its mass. This in turn assumes we know how to define force, but that is not so hard. A simple but not very accurate or practical way would be to measure force in terms of the pull exerted by some standard thing; an average tortoise, let us say, walking in a straight line with a harness attaching it to the object being pulled. We could term the average tortoise the 'SI Tortoise' and keep it in a sealed box in the International Bureau of Weights and Measures in Sèvres, France. Two harnessed tortoises would exert twice the force, three would exert three times the force and so on. We could then always talk about any push or pull in terms of the number of average tortoises required to generate it.

Given this system, which is ridiculous enough to be agreed on by any international committee of standards,[1] we can simply pull an object with a tortoise and measure its acceleration, and this will allow us to deduce its mass using Newton's second law. We can then repeat the process for a second object to deduce its mass and then we can put both masses into the law of gravity to determine the force between the masses due to gravity. To put a tortoise-equivalent

---

1. But not so ridiculous when you consider that an oft-used unit of power, even to this day, is the 'horsepower'.

number on the gravitational force between two masses, though, we would still need to calibrate the whole system to the strength of gravity itself, and this is where the symbol G comes in.

G is a very important number, called 'Newton's gravitational constant', which encodes the strength of the gravitational force. If we doubled G, we would double the force, and this would make the apple accelerate at double the rate towards the ground. It therefore describes one of the fundamental properties of our Universe and we would live in a very different Universe if it took on a different value. It is currently thought that G takes the same value everywhere in the Universe, and that it has remained constant throughout all of time (it appears in Einstein's theory of gravity too, where it is also a constant). There are other universal constants of Nature that we'll meet in this book. In quantum mechanics, the most important is Planck's constant, named after quantum pioneer Max Planck and given the symbol h. We shall also need the speed of light, c, which is not only the speed that light travels in a vacuum but the universal speed limit. 'It is impossible to travel faster than the speed of light and certainly not desirable,' Woody Allen once said, 'as one's hat keeps blowing off.'

Newton's three laws of motion and the law of gravitation are all that is needed to understand motion in the presence of gravity. There are no other hidden rules that we did not state – just these few laws do the trick and allow us, for example, to understand the orbits of the planets in our solar system. Together, they severely restrict the sort of paths that objects are allowed to take when moving under the influence of gravity. It can be proved using only Newton's laws that all of the planets, comets, asteroids and meteors in our solar system are only allowed to move along paths known as conic sections. The simplest of these, and the one that the Earth follows to a very good approximation in its orbit around the Sun, is a circle. More generally, planets and moons move along orbital paths known as ellipses, which are like stretched circles. The other two conic sections are known as the parabola and the hyperbola. A parabola is the path that a cannonball takes when fired from the cannon. The final conic section, the hyperbola, is the path that the most distant object ever constructed by human kind is now following outwards to the stars. Voyager 1 is, at the time of writing, around 17,610,000,000 km from the Earth, and travelling away from the solar system at a speed of 538,000,000 km per year. This most beautiful of engineering achievements was launched in 1977 and is still in contact with the Earth, recording measurements of the solar wind on a tape recorder and transmitting them back with a power of 20 watts. Voyager 1, and her sister ship Voyager 2, are inspiring testaments to the human desire to explore our Universe. Both spacecraft visited Jupiter and Saturn and Voyager 2 went on to visit Uranus and Neptune. They navigated the solar system with precision, using gravity to slingshot them beyond the planets and into interstellar space. Navigators here on Earth used nothing more than Newton's laws to plot their courses between the inner and outer planets and outwards to the stars. Voyager 2 will sail close to Sirius, the brightest star in the skies, in just under 300,000 years. We did all this, and we know all this, because of Newton's theory of gravity and his laws of motion.

Newton's laws provide us with a very intuitive picture of the world. As we have seen, they take the form of equations – mathematical relationships between measurable quantities – that allow us to predict with precision how objects move around. Inherent in the whole framework is the assumption that objects are, at any instant, located somewhere and that, as time passes, objects move smoothly around from place to place. This seems so self-evidently true that it is hardly worth commenting upon, but we need to recognize that this is a prejudice. Can we really be sure that things are definitely here or there, and that they are not actually in two different places at the same time? Of course, your garden shed is not in any noticeable sense sitting in two distinctly different places at once – but how about an electron in an atom? Could that be both 'here' and 'there'? Right now that kind of suggestion sounds crazy, mainly because we can't picture it in our mind's eye, but it will turn out to be the way things actually work. At this stage in our narrative, all we are doing in making this strange-sounding statement is pointing out that Newton's laws are built on intuition, and that is like a house built on sand as far as fundamental physics is concerned.

There is a very simple experiment, first conducted by Clinton Davisson and Lester Germer at Bell Laboratories in the United States and published in 1927, which shows that Newton's intuitive picture of the world is wrong. Although apples, planets and people certainly appear to behave in a 'Newtonian' way, gliding from place to place in a regular and predictable fashion as time unfolds, their experiment showed that the fundamental building blocks of matter do not behave at all like this.

Davisson and Germer's paper begins: 'The intensity of scattering of a homogeneous beam of electrons of adjustable speed incident upon a single crystal of nickel has been measured as a function of direction.' Fortunately, there is a way to appreciate the key content of their findings using a simplified version of their experiment, known as the double-slit experiment. The experiment consists of a source that sends electrons towards a barrier with two small slits (or holes) cut into it. On the other side of the barrier, there is a screen that glows when an electron hits it. It doesn't matter what the source of electrons is, but practically speaking one can imagine a length of hot wire stretched out along the side of the experiment.[2] We've sketched the double-slit experiment in Figure 2.2.

Imagine pointing a camera at the screen and leaving the shutter open to take a long-exposure photograph of the little flashes of light emitted as, one by one, the electrons hit it. A pattern will build up, and the simple question is, what is the pattern? Assuming electrons are simply little particles that behave rather like apples or planets, we might expect the emergent pattern to look something like that shown in Figure 2.2. Some electrons go through the slits, most don't. The ones that make it through might bounce off the

---

2. Once upon a time, televisions operated using this idea. A stream of electrons generated by a hot wire was gathered, focused into a beam and deflected by a magnetic field onto a screen that glowed when the electrons hit it.

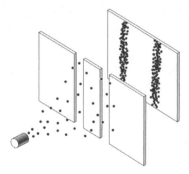

Figure 2.2: An electron-gun source fires electrons towards a pair of slits and, if the electrons behaved like 'regular' particles, we would expect to see hits on the screen that build up a pair of stripes, as illustrated. Remarkably, this is *not* what happens.

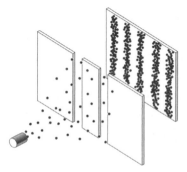

Figure 2.3: In reality the electrons do not hit the screen aligned with the slits. Instead they form a stripy pattern: electron by electron, the stripes build up over time.

edge of the slits a bit, which will spread them out, but the most hits, and therefore the brightest bits of the photograph, will surely appear directly aligned with the two slits.

This isn't what happens. Instead, the picture looks like Figure 2.3. A pattern like this is what Davisson and Germer published in their 1927 paper. Davisson subsequently received the 1937 Nobel Prize for the 'experimental discovery of electron diffraction by crystals'. He shared the prize, not with Germer, but with George Paget Thomson, who saw the same pattern independently in experiments at the University of Aberdeen. The alternating stripes of light and dark are known as an interference pattern, and interference is more usually associated with waves. To understand why, let's imagine doing the double-slit experiment with water waves instead of electrons.

Imagine a water tank with a wall midway down with two slits cut into it. The screen and camera could be replaced with a wave-height detector, and the hot wire with something that makes waves: a plank of wood along the side of the tank attached to a motor that keeps it dipping in and out of the water would do. The waves from the plank will travel across the surface of the water until they reach the wall. When a wave hits the wall, most of it will bounce back, but two small pieces will pass through the slits. These two new waves will spread outwards from the slits towards the wave-height detector. Notice that we used the term 'spread out' here, because the waves don't just carry on in a straight line from the slits. Instead, the slits act as two sources of new waves, each issuing forth in ever increasing semi-circles. Figure 2.4 illustrates what happens.

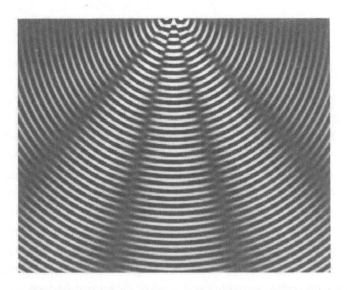

Figure 2.4. An aerial view of water waves emanating from two points in a tank of water (they are located at the top of the picture). The two circular waves overlap and interfere with each other. The 'spokes' are the regions where the two waves have cancelled each other out and the water there remains undisturbed.

The figure provides a striking visual demonstration of the behaviour of waves in water. There are regions where there are no waves at all, which seem to radiate out from the slits like the spokes of a wheel, whilst other regions are still filled with the peaks and troughs of the waves. The parallels with the pattern seen by Davisson, Germer and Thomson are striking. For the case of electrons hitting the screen, the regions where few electrons are detected correspond to the places in the tank where the water surface remains flat – the spokes you can see radiating outwards in the figure.

In a tank of water it is quite easy to understand how these spokes emerge: it is in the mixing and merging of the waves as they spread out from the slits. Because waves have peaks and troughs, when two waves meet they can either add or subtract. If two waves meet such that the peak of one is aligned with the trough of the other, they will cancel out and there will be no wave at that point. At a different place, the waves might arrive with their peaks in perfect alignment, and here they will add to produce a bigger wave. At each point in the water tank, the distance between it and the two slits will be a little different, which means that at some places the two waves will arrive with peaks together, at others with peaks and troughs aligned and, in most places, with some combination of these two extremes. The result will be an alternating pattern; an interference pattern.

In contrast to water waves, the experimentally observed fact that electrons also produce an interference pattern is very difficult to understand. According to Newton and common sense, the electrons emerge from the source, travel in straight lines towards the slits (because there are no forces acting on them – remember Newton's first law), pass through with perhaps a slight deflection if they glance off the edge of the slit, and continue in a straight line until they hit the screen. But this would not result in an interference pattern – it would give the pair of stripes as shown in Figure 2.2. Now we could suppose that there is some clever mechanism whereby the electrons exert a force on each other so as to deflect themselves from straight lines as they stream through the slits. But this can be ruled out because we can set the experiment up such that we send just one electron at a time from source to screen. You would have to wait, but, slowly and surely, as the electrons hit the screen one after the other, the stripy pattern would build up. This is very surprising because the stripy pattern is absolutely characteristic of waves interfering with each other, yet it emerges one electron at a time – dot by dot. It's a good mental exercise to try to imagine how it could be that, particle by particle, an interference pattern builds up as we fire tiny bullet-like particles at a pair of slits in a screen. It's a good exercise because it's futile, and a few hours of brain racking should convince you that a stripy pattern is inconceivable. Whatever those particles are that hit the screen, they are not behaving like 'regular' particles. It is as if the electrons are in some sense 'interfering with themselves'. The challenge for us is to come up with a theory that can explain what that means.

There is an interesting historical coda to this story, which provides a glimpse into the intellectual challenge raised by the double-slit experiment. George Paget Thomson was the son of J. J. Thomson, who himself received a Nobel Prize for his discovery of the electron in 1899. J. J. Thomson showed that the electron is a particle, with a particular electric charge and a particular mass; a tiny, point-like grain of matter. His son received the Nobel Prize forty years later for showing that the electron doesn't behave as his father might have expected. Thomson senior was not wrong; the electron does have a well-defined mass and electric charge, and every time we see one it appears as a little point of matter. It just doesn't seem to behave *exactly* like a regular particle, as Davisson, Germer and Thomson junior discovered. Importantly, though, it doesn't behave *exactly* like a wave either because the pattern is not built up as a result of some smooth deposition of energy; rather it is built out of many tiny dots. We always detect Thomson senior's single, point-like electrons.

Perhaps you can already see the need to engage with Heisenberg's way of thinking. The things we observe are particles, so we had better construct a theory of particles. Our theory must also be able to predict the appearance of the interference pattern that builds up as the electrons, one after another, pass through the slits and hit the screen. The details of how the electrons travel from source to slits to screen are not something we observe, and therefore need not be in accord with anything we experience in daily life. Indeed, the electron's 'journey' need not even be something we can talk about at all. All we have to do is find a theory capable of predicting that the electrons hit the screen in the pattern observed in the double-slit experiment. This is what we will do in the next chapter.

Lest we lapse into thinking that this is merely a fascinating piece of micro-physics that has little relevance to the world at large, we should say that the quantum theory of particles we develop to explain the double-slit experiment will also turn out to be capable of explaining the stability of atoms, the coloured light emitted from the chemical elements, radioactive decay, and indeed all of the great puzzles that perplexed scientists at the turn of the twentieth century. The fact that our framework describes the way electrons behave when locked away inside matter will also allow us to understand the workings of quite possibly the most important invention of the twentieth century: the transistor.

In the very final chapter of this book, we will meet a striking application of quantum theory that is one of the great demonstrations of the power of scientific reasoning. The more outlandish predictions of quantum theory usually manifest themselves in the behaviour of small things. But, because large things are made of small things, there are certain circumstances in which quantum physics is required to explain the observed properties of some of the most massive objects in the Universe – the stars. Our Sun is fighting a constant battle with gravity. This ball of gas a third of a million times more massive than our planet has a gravitational force at its surface that is almost twenty-eight times that at the Earth, which provides a powerful incentive for it to collapse in on itself. The collapse is prevented by the outward pressure generated by nuclear fusion reactions deep within the solar core as 600 million tonnes of hydrogen are converted into helium every second. Vast though our star is, burning fuel at such a ferocious rate must ultimately have consequences, and one day the Sun's fuel source will run out. The outward pressure will then cease and the force of gravity will reassert its grip unopposed. It would seem that nothing in Nature could stop a catastrophic collapse.

In reality, quantum physics steps in and saves the day. Stars that have been rescued by quantum effects in this way are known as white dwarves, and such will be the final fate of our Sun. At the end of this book we will employ our understanding of quantum mechanics to determine the maximum mass of a white dwarf star. This was first calculated, in 1930, by the Indian astrophysicist Subrahmanyan Chandrasekhar, and it turns out to be approximately 1.4 times the mass of our Sun. Quite wonderfully, that number can be computed using only the mass of a proton and the values of the three constants of Nature we have already met: Newton's gravitational constant, the speed of light, and Planck's constant.

The development of the quantum theory itself and the measurement of these four numbers could conceivably have been achieved without ever looking at the stars. It is possible to imagine a particularly agoraphobic civilization confined to deep caves below the surface of their home planet. They would have no concept of a sky, but they could have developed quantum theory. Just for fun, they may even decide to calculate the maximum mass of a giant sphere of gas. Imagine that, one day, an intrepid explorer chooses to venture above ground for the first time and gaze in awe at the spectacle above: a sky full of lights; a galaxy of a hundred billion suns arcing from horizon to horizon. The explorer would find, just as we have found from our vantage point here on Earth, that out there amongst the many fading remnants of dying stars there is not a single one with a mass exceeding the Chandrasekhar limit.

3. What Is a Particle?

Our approach to quantum theory was pioneered by Richard Feynman, the Nobel Prize-winning, bongo-playing New Yorker described by his friend and collaborator Freeman Dyson as 'half genius, half buffoon'. Dyson later changed his opinion: Feynman could be more accurately described as 'all genius, all buffoon'. We will follow his approach in our book because it is fun, and probably the simplest route to understanding our Quantum Universe.

As well as being responsible for the simplest formulation of quantum mechanics, Richard Feynman was also a great teacher, able to transfer his deep understanding of physics to the page or lecture theatre with unmatched clarity and a minimum of fuss. His style was contemptuous of those who might seek to make physics more complicated than it need be. Even so, at the beginning of his classic undergraduate textbook series *The Feynman Lectures on Physics*, he felt the need to be perfectly honest about the counterintuitive nature of the quantum theory. Subatomic particles, Feynman wrote, 'do not behave like waves, they do not behave like particles, they do not behave like clouds, or billiard balls, or weights on springs, or like anything that you have ever seen'. Let's get on with building a model for exactly how they do behave.

As our starting point we will assume that the elemental building blocks of Nature are particles. This has been confirmed not only by the double-slit experiment, where the electrons always arrive at specific places on the screen, but by a whole host of other experiments. Indeed 'particle physics' is not called that for nothing. The question we need to address is: how do particles move around? Of course, the simplest assumption would be that they move in nice straight lines, or curved lines when acted upon by forces, as dictated by Newton. This cannot be correct though, because any explanation of the double-slit experiment requires that the electrons 'interfere with themselves' when they pass through the slits, and to do that they must in some sense be spread out. This therefore is the challenge: build a theory of point-like particles such that those same particles are also spread out. This is not as impossible as it sounds: we can do it if we let any single particle be *in many places at once*. Of course, that may still sound impossible, but the proposition that a particle should be in many places at once is actually a rather clear statement, even if it sounds silly. From now on, we'll refer to these counterintuitive, spread-out-yet-point-like particles as quantum particles.

With this 'a particle can be in more than one place at once' proposal, we are moving away from our everyday experience and into uncharted territory. One of the major obstacles to developing an understanding of quantum physics is the confusion this kind of thinking can engender. To avoid confusion, we should follow Heisenberg and learn to feel comfortable with views of the world that run counter to tangible experience. Feeling 'uncomfortable' can be mistaken for 'confusion', and very often students of quantum physics continue to attempt to understand what is happening in everyday terms. It is the resistance to new ideas that actually leads to confusion, not the inherent difficulty of the ideas themselves, because the real world simply doesn't behave in an everyday way. We must therefore keep an open mind and not be distressed by all the weirdness. Shakespeare had it right when Hamlet says, 'And therefore as a stranger give it welcome. There are more things in heaven and earth, Horatio, Than are dreamt of in your philosophy.'

A good way to begin is to think carefully about the double-slit experiment for water waves. Our aim will be to work out just what it is about waves that causes the interference pattern. We should then make sure that our theory of quantum particles is capable of encapsulating this behaviour, so that we can have a chance of explaining the double-slit experiment for electrons.

There are two reasons why waves journeying through two slits can interfere with themselves. The first is that the wave travels through *both of the slits at once*, creating two new waves that head off and mix together. It's obvious that a wave can do this. We have no problem visualizing one long, ocean wave rolling to the shore and crashing on to a beach. It is a wall of water; an extended, travelling thing. We are therefore going to need to decide how to make our quantum particle 'an extended, travelling thing'. The second reason is that the two new waves heading out from the slits are able either to add or to subtract from each other when they mix. This ability for two waves to interfere is clearly crucial in explaining the interference pattern. The extreme case is when the peak of one wave coincides with the trough of another, in which case they completely cancel each other out. So we are also going to need to allow our quantum particle to interfere somehow with itself.

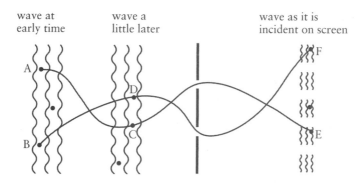

Figure 3.1. How the wave describing an electron moves from source to screen and how it should be interpreted as representing all of the ways that the electron travels. The paths A to C to E and B to D to F illustrate just two of the infinity of possible paths the single electron does take.

The double-slit experiment connects the behaviour of electrons and the behaviour of waves, so let us see how far we can push the connection. Take a look at Figure 3.1 and, for the time being, ignore the lines joining A to E and B to F and concentrate on the waves. The figure could then describe a water tank, with the wavy lines representing, from left to right, how a water wave rolls its way across the tank. Imagine taking a photograph of the tank just after a plank of wood has splashed in on the left-hand side to make a wave. The snapshot would reveal a newly formed wave that extends from top to bottom in the picture. All the water in the rest of the tank would be calm. A second snapshot taken a little later reveals that the water wave has moved towards the slits, leaving flat water behind it. Later still, the water wave passes through the pair of slits and generates the stripy interference pattern illustrated by the wavy lines on the far right.

Now let us reread that last paragraph but replace 'water wave' with 'electron wave', whatever that may mean. An electron wave, suitably interpreted, has the potential to explain the stripy pattern we want to understand as it rolls through the experiment like a water wave. But we do need to explain why the electron pattern is made up of tiny dots as the electrons hit the screen one by one. At first sight that seems in conflict with the idea of a smooth wave, but it is not. The clever bit is to realize that we can offer an explanation if we interpret the electron wave not as a real material disturbance (as is the case with a water wave), but rather as something that simply informs us where the electron is likely to be found. Notice we said 'the' electron because the wave is to describe the behaviour of a single electron – that way we have a chance of explaining how those dots emerge. This is an electron wave, and not a wave of electrons: we must never fall into the trap of thinking otherwise. If we imagine a snapshot of the wave at some instant in time, then we want to interpret it such that where the wave is largest the electron is most likely to be found, and where the wave is smallest the electron is least likely to be found. When the wave finally reaches the screen, a little spot appears and informs us of the location of the electron. The sole job of the electron wave is to allow us to compute the odds that the electron hits the screen at some particular place. If we do not worry about what the electron wave actually 'is', then everything is straightforward because once we know the wave then we can say where the electron is likely to be. The fun comes next, when we try to understand what this proposal for an electron wave implies for the electron's journey from slit to screen.

Before we do this, it might be worth reading the above paragraph again because it is very important. It's not supposed to be obvious and it is certainly not intuitive. The 'electron wave' proposal has all the necessary properties to explain the appearance of the experimentally observed interference pattern, but it is something of a guess as to how things might work out. As good physicists we should work out the consequences and see if they correspond to Nature.

Returning to Figure 3.1, we have proposed that at each instant in time the electron is described by a wave, just as in the case of water waves. At an early time, the electron wave is to the left of the slits. This means that the electron is in some sense located somewhere within the wave. At a later time, the wave will advance towards the slits just as the water wave did, and the electron will now be somewhere in the new wave. We are saying that the electron 'could be first at A and then at C', or it 'could be first at B and then at D', or it 'could be at A and then at D', and so on. Hold that thought for a minute, and think about an even later time, after the wave has passed through the slits and reached the screen. The electron could now be found at E or perhaps at F. The curves that we have drawn on the diagram represent two possible paths that the electron could have taken from the source, through the slits and onto the screen. It could have gone from A to C to E, and it could have gone from B to D to F. These are just two out of an infinite number of possible paths that the electron could have taken.

The crucial point is that it makes no sense to say that 'the electron could have ventured along each of these routes, but really it went along only one of them'. To say that the electron really ventured along one particular path would be to give ourselves no more of a chance of explaining the interference pattern than if we had blocked up one of the slits in the water wave experiment. We need to allow the wave to go through both slits in order to get an interference pattern, and this means that we must allow all the possible paths for the electron to travel from source to screen. Put another way, when we said that the electron is 'somewhere within the wave' we really meant to say that it is simultaneously everywhere in the wave! This is how we must think because if we suppose the electron is actually located at some specific point, then the wave is no longer spread out and we lose the water wave analogy. As a result, we cannot explain the interference pattern.

Again, it might be worth rereading the above piece of reasoning because it motivates much of what follows. There is no sleight of hand: what we are saying is that we need to describe a spread-out wave that is also a point-like electron, and one possible way to achieve this is to say that the electron sweeps from source to screen following all possible paths at once.

This suggests that we should interpret an electron wave as describing a single electron that travels from source to screen by an infinity of different routes. In other words, the correct answer to the question 'how did that electron get to the screen' is 'it travelled by an infinity of possible routes, some of which went through the upper slit and some of which went though the lower one'. Clearly the 'it' that is the electron is not an ordinary, everyday particle. This is what it means to be a quantum particle.

Having decided to seek a description of an electron that mimics in many ways the behaviour of waves, we need to develop a more precise way to talk about waves. We shall begin with a description of what is happening in a water tank when two waves meet, mix and interfere with each other. To do this, we must find a convenient way of representing the positions of the peaks and troughs of each wave. In the technical jargon, these are known as phases. Colloquially things are described as 'in phase' if they reinforce one another in some way, or 'out of phase' if they cancel each other out. The word is also used to describe the Moon: over the course of 29.5 days, the Moon passes from new to full and back again in a continuous waxing and waning cycle. The etymology of the word 'phase' stems from the Greek *phasis*, which means the appearance and disappearance of an astronomical phenomenon, and the regular appearance and disappearance of the bright lunar surface

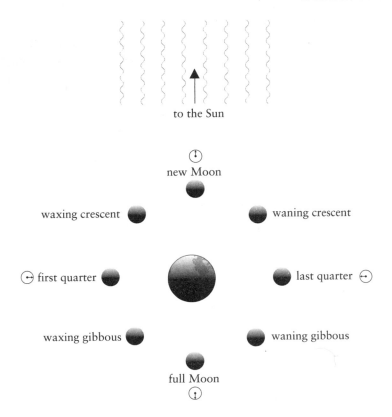

to the Sun

new Moon

waxing crescent   waning crescent

first quarter   last quarter

waxing gibbous   waning gibbous

full Moon

Figure 3.2. The phases of the Moon.

seems to have led to its twentieth-century usage, particularly in science, as a description of something cyclical. And this is a clue as to how we might find a pictorial representation of the positions of the peaks and troughs of water waves.

Have a look at Figure 3.2. One way to represent a phase is as a clock face with a single hand rotating around. This gives us the freedom to represent visually a full 360 degrees worth of possibilities: the clock hand can point to 12 o'clock, 3 o'clock, 9 o'clock and all points in between. In the case of the Moon, you could imagine a new Moon represented by a clock hand pointing to 12 o'clock, a waxing crescent at 1:30, the first quarter at 3, the waxing gibbous at 4:30, the full Moon at 6 and so on. What we are doing here is using something abstract to describe something concrete; a clock face to describe the phases of the Moon. In this way we could draw a clock with its hand pointing to 12 o'clock and you'd immediately know that the clock represented a new Moon. And even though we haven't actually said it, you'd know that a clock with the hand pointing to 5 o'clock would mean that we are approaching a full Moon. The use of abstract pictures or symbols to represent real things is absolutely fundamental in physics – this is essentially what physicists use mathematics for. The power of the approach comes when the abstract pictures can be manipulated using simple rules to make firm predictions about the real world. As we'll see in a moment, the clock faces will allow us to do just this because they are able to keep track of the relative positions of the peaks and troughs of waves. This in turn will allow us to calculate whether they will cancel or reinforce one another when they meet.

Figure 3.3 shows a sketch of two water waves at an instant in time. Let's represent the peaks of the waves by clocks reading 12 o'clock and the troughs by clocks reading 6 o'clock. We can also represent places on the waves intermediate between peaks and troughs with clocks reading intermediate times, just as we did for the phases of the Moon between new and full. The distance between the successive peaks and troughs of the wave is an important number; it is known as the wavelength of the wave.

The two waves in Figure 3.3 are out of phase with each other, which means that the peaks of the top wave are aligned with the troughs of the bottom wave, and vice versa. As a result it is pretty clear that they will entirely cancel each other out when we add them together. This is illustrated at the bottom of the figure, where the 'wave' is flat-lining. In terms of clocks, all of the 12 o'clock clocks for the top wave, representing its peaks, are aligned with the 6 o'clock clocks for the bottom wave, representing its troughs. In fact, everywhere you look, the clocks for the top wave are pointing in the opposite direction to the clocks for the bottom wave.

Using clocks to describe waves does, at this stage, seem like we

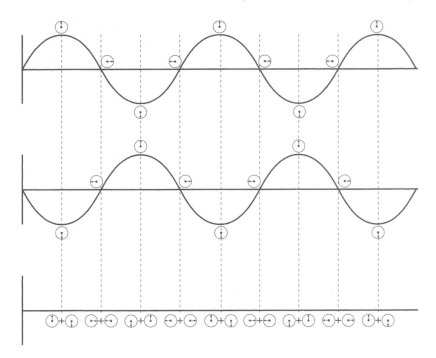

Figure 3.3. Two waves arranged such that they cancel out completely. The top wave is out of phase with the second wave, i.e. peaks align with troughs. When the two waves are added together they cancel out to produce nothing – as illustrated at the bottom where the 'wave' is flat-lining.

are over-complicating matters. Surely if we want to add together two water waves, then all we need to do is add the heights of each of the waves and we don't need clocks at all. This is certainly true for water waves, but we are not being perverse and we have introduced the clocks for a very good reason. We will discover soon enough that the extra flexibility they allow is absolutely necessary when we come to use them to describe quantum particles.

With this in mind, we shall now spend a little time inventing a precise rule for adding clocks. In the case of Figure 3.3, the rule must result in all the clocks 'cancelling out', leaving nothing behind: 12 o'clock cancels 6 o'clock, 3 o'clock cancels 9 o'clock and so on. This perfect cancellation is, of course, for the special case when the waves are perfectly out of phase. Let's search for a general rule that will work for the addition of waves of any alignment and shape.

Figure 3.4 shows two more waves, this time aligned in a different way, such that one is only slightly offset against the other. Again, we have labelled the peaks, troughs and points in between with clocks. Now, the 12 o'clock clock of the top wave is aligned with the 3 o'clock clock of the bottom wave. We are going to state a rule that allows us to add these two clocks together. The rule is that we take the two hands and stick them together head to tail. We then complete the triangle by drawing a new hand joining the other two hands together. We have sketched this recipe in Figure 3.5. The new hand will be a different length to the other two, and point in a different direction; it is a new clock face, which is the sum of the other two.

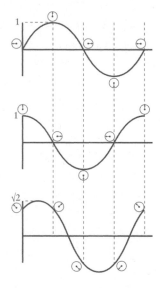

Figure 3.4. Two waves offset relative to each other. The top and middle waves add together to produce the bottom wave.

We can be more precise now and use simple trigonometry to calculate the effect of adding together any specific pair of clocks. In Figure 3.5 we are adding together the 12 o'clock and 3 o'clock clocks. Let's suppose that the original clock hands are of length 1 cm (corresponding to water waves of peak height equal to 1 cm). When we place the hands head-to-tail we have a right-angled triangle with two sides each of length 1 cm. The new clock hand will be the length of the third side of the triangle: the hypotenuse. Pythagoras' Theorem tells us that the square of the hypotenuse is equal to the sum of the squares of the other two sides: $h^2 = x^2 + y^2$. Putting the numbers in, $h^2 = 1^2 + 1^2 = 2$. So the length of the new clock hand h is the square root of 2, which is approximately 1.414 cm. In what direction will the new hand point? For this we need to know the angle in our triangle, labelled θ in the figure. For the particular example of two hands of equal length, one pointing to 12 o'clock and one to 3 o'clock, you can probably work it out without knowing any trigonometry at all. The hypotenuse obviously points at an angle of 45 degrees, so the new 'time' is half way between 12 o'clock and 3 o'clock, which is half past one. This example is a special case, of course. We chose the clocks so that the hands were at right angles and of the same length to make the mathematics easy. But it is obviously possible to work out the length of the hand and time resulting from the addition of any pair of clock faces.

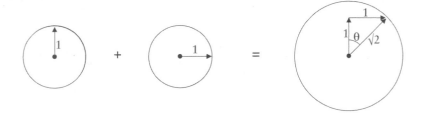

Figure 3.5. The rule for adding clocks.

Now look again at Figure 3.4. At every point along the new wave, we can compute the wave height by adding the clocks together using the recipe we just outlined and asking how much of the new clock hand points in the 12 o'clock direction. When the clock points to 12 o'clock this is obvious – the height of the wave is simply the length of the clock hand. Similarly at 6 o'clock, it's obvious because the wave has a trough with a depth equal to the length of the hand.

It's also pretty obvious when the clock reads 3 o'clock (or 9 o'clock) because then the wave height is zero, since the clock hand is at right angles to the 12 o'clock direction. To compute the wave height described by any particular clock we should multiply the length of the hand, h, by the cosine of the angle the hand makes with the 12 o'clock direction. For example, the angle that a 3 o'clock makes with 12 o'clock is 90 degrees and the cosine of 90 degrees is zero, which means the wave height is zero. Similarly, a time of half-past-one corresponds to an angle of 45 degrees with the 12 o'clock direction and the cosine of 45 degrees is approximately 0.707, so the height of the wave is 0.707 times the length of the hand (notice that 0.707 is $1/\sqrt{2}$). If your trigonometry is not up to those last few sentences then you can safely ignore the details. It's the principle that matters, which is that, given the length of a clock hand and its direction you can go ahead and calculate the wave height – and even if you don't understand trigonometry you could make a good stab at it by carefully drawing the clock hands and projecting on to the 12 o'clock direction using a ruler. (We would like to make it very clear to any students reading this book that we do not recommend this course of action: sines and cosines are useful things to understand.)

That's the rule for adding clocks, and it works a treat, as illustrated in the bottom of the three pictures in Figure 3.4, where we have repeatedly applied the rule for various points along the waves.

In this description of water waves, all that ever matters is the projection of the 'time' in the 12 o'clock direction, corresponding to

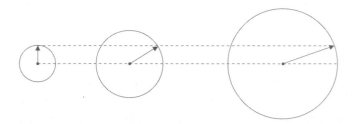

Figure 3.6. Three different clocks all with the same projection in the 12 o'clock direction.

just one number: the wave height. That is why the use of clocks is not really necessary when it comes to describing water waves. Take a look at the three clocks in Figure 3.6: they all correspond to the same wave height and so they provide equivalent ways of representing the same height of water. But clearly they are different clocks and, as we shall see, these differences do matter when we come to use them to describe quantum particles because, for them, the length of the clock hand (or equivalently the size of the clock) has a very important interpretation.

At some points in this book and at this point especially, things are abstract. To keep ourselves from succumbing to dizzying confusion, we should remember the bigger picture. The experimental results of Davisson, Germer and Thomson, and their similarity with the behaviour of water waves, have inspired us to make an ansatz: we should represent a particle by a wave, and the wave itself can be represented by lots of clocks. We imagine that the electron wave propagates 'like a water wave', but we haven't explained how that works in any detail. But then we never said how the water wave propagates either. All that matters for the moment is that we recognize the analogy with water waves, and the notion that the electron is described at any instant by a wave that propagates and interferes like water waves do. In the next chapter we will do better than this and be more precise about how an electron actually moves around as time unfolds. In doing that we will be led to a host of treasures, including Heisenberg's famous Uncertainty Principle.

Before we move on to that, we want to spend a little time talking about the clocks that we are proposing to represent the electron wave. We emphasize that these clocks are not real in any sense, and their hour hand has absolutely nothing to do with what time of day it is. This idea of using an array of little clocks to describe a real physical phenomenon is not so bizarre a concept as it may seem. Physicists use similar techniques to describe many things in Nature, and we have already seen how they can be used to describe water waves.

Another example of this type of abstraction is the description of the temperature in a room, which can be represented using an array of numbers. The numbers do not exist as physical objects any more than our clocks do. Instead, the set of numbers and their association with points in the room is simply a convenient way of representing the temperature. Physicists call this mathematical structure a field. The temperature field is simply an array of numbers, one for every point. In the case of a quantum particle, the field is more complicated because it requires a clock face at each point rather than a single number. This field is usually called the wavefunction of the particle. The fact that we need an array of clocks for the wavefunction, whilst a single number would suffice for the temperature field or for water waves, is an important difference. In physics jargon, the clocks are there because the wavefunction is a 'complex' field, whilst the temperature or water wave heights are both 'real' fields. We shall not need any of this language, because we can work with the clock faces.[1]

We should not worry that we have no direct way to sense a wavefunction, in contrast to a temperature field. The fact that it is not something we can touch, smell or see directly is irrelevant. Indeed, we would not get very far in physics if we decided to restrict our description of the Universe to things we can directly sense.

In our discussion of the double-slit experiment for electrons, we said that the electron wave is largest where the electron is most likely to be. This interpretation allowed us to appreciate how the stripy interference pattern can be built up dot by dot as the electrons arrive. But this is not a precise enough statement for our purposes now. We want to know what the probability is to find an electron at a particular point – we want to put a number on it. This is where the clocks become necessary, because the probability that we want is not simply the wave height. The correct thing to do is to

---

1. For those who are familiar with mathematics, just exchange the words as follows: 'clock' for 'complex number', 'size of the clock' for 'modulus of the complex number' and 'the direction of the hour-hand' for 'the phase'. The rule for adding clocks is nothing more than the rule for adding complex numbers.

interpret the *square* of the length of the clock hand as the probability to find the particle at the site of the clock. This is why we need the extra flexibility that the clocks give us over simple numbers. That interpretation is not meant to be at all obvious, and we cannot offer any good explanation for why it is correct. In the end, we know that it is correct because it leads to predictions that agree with experimental data. This interpretation of the wavefunction was one of the thorny issues facing the early pioneers of quantum theory.

The wavefunction (that is our cluster of clocks) was introduced into quantum theory in a series of papers published in 1926 by the Austrian physicist Erwin Schrödinger. His paper of 21 June contains an equation that should be etched into the mind of every undergraduate physics student. It is known, naturally enough, as the Schrödinger equation:

$$i\hbar \frac{\partial}{\partial t} \Psi = \hat{H} \Psi$$

The Greek symbol Ψ (pronounced 'psi') represents the wavefunction, and the Schrödinger equation describes how it changes as time passes. The details of the equation are irrelevant for our purposes because we are not going to follow the Schrödinger approach in this book. What is interesting, though, is that, although Schrödinger wrote down the correct equation for the wavefunction, he initially got the interpretation wrong. It was Max Born, one of the oldest of the physicists working on the quantum theory in 1926, who, at the grand old age of forty-three, gave the correct interpretation in a paper submitted just four days after Schrödinger's. We mention his age because quantum theory during the mid 1920s gained the nickname 'Knabenphysik' – 'boy physics' – because so many of the key protagonists were young. In 1925 Heisenberg was twenty-three, Wolfgang Pauli, whose famous Exclusion Principle we shall meet later on, was twenty-two, as was Paul Dirac, the British physicist who first wrote down the correct equation describing the electron. It is often claimed that their youth freed them from the old ways of thinking and allowed them fully to embrace the radical new picture of the world represented by quantum theory. Schrödinger, at thirty-eight, was an old man in this company and it is true that he was never completely at ease with the theory he played such a key role in developing.

Born's radical interpretation of the wavefunction, for which he received the Nobel Prize for physics in 1954, was that the square of the length of the clock hand at a particular point represents the probability of finding a particle there. For example, if the hour-hand on the clock located at some place has a length of 0.1 then squaring this gives 0.01. This means that the probability to find the particle at this place is 0.01, i.e. one in a hundred. You might ask why Born didn't just square the clocks up in the first place, so that in the last example the clock hand would itself have a length of 0.01. That will not work, because to account for interference we are going to want to add clocks together and adding 0.01 to 0.01 say (which gives 0.02) is not the same as adding 0.1 to 0.1 and then squaring (which gives 0.04).

We can illustrate this key idea in quantum theory with another example. Imagine doing something to a particle such that it is described by a specific array of clocks. Also imagine we have a device that can measure the location of particles. A simple-to-imagine-but-not-so-simple-to-build device might be a little box that we can rapidly erect around any region of space. If the theory says that the chance of finding a particle at some point is 0.01 (because the clock hand at that point has length 0.1), then when we erect the box around that point we have a one in a hundred chance of finding the particle inside the box afterwards. This means that it is unlikely that we'll find anything in the box. However, if we are able to reset the experiment by setting everything up such that the particle is once again described by the same initial set of clocks, then we could redo the experiment as many times as we wish. Now, for every 100 times we look in the little box we should, on average, discover that there is a particle inside it once – it will be empty the remaining ninety-nine times.

The interpretation of the squared length of the clock hand as the probability to find a particle at a particular place is not particularly difficult to grasp, but it does seem as if we (or to be more precise, Max Born) plucked it out of the blue. And indeed, from a historical perspective, it proved very difficult for some great scientists, Einstein and Schrödinger among them, to accept. Looking back on the summer of 1926, fifty years later, Dirac wrote: 'This problem of getting the interpretation proved to be rather more difficult than just working out the equations.' Despite this difficulty, it is noteworthy that by the end of 1926 the spectrum of light emitted from the hydrogen atom, one of the great puzzles of nineteenth-century physics, had already been computed using both Heisenberg's and Schrödinger's equations (Dirac eventually proved that their two approaches were in all cases entirely equivalent).

Einstein famously expressed his objection to the probabilistic nature of quantum mechanics in a letter to Born in December 1926. 'The theory says a lot but does not really bring us any closer to the secret of the "old one". I, at any rate, am convinced that *He* is not playing at dice.' The issue was that, until then, it had been assumed that physics was completely deterministic. Of course, the idea of probability is not exclusive to quantum theory. It is regularly used in a variety of situations, from gambling on horse races to the science of thermodynamics, upon which whole swathes of Victorian engineering rested. But the reason for this is a lack of knowledge about the part of the world in question, rather than something fundamental. Think about tossing a coin – the archetypal game of chance. We are all familiar with probability in this context. If we toss the coin 100 times, we expect, on average, that fifty times it will land heads and fifty times tails. Pre-quantum theory, we were obliged to say that, if we knew everything there is to know about the coin – the precise way we tossed it into the air, the pull of gravity, the details of little air currents that swish through the room, the temperature of the air, etc. – then we could, *in principle*, work out whether the coin would land heads or tails. The emergence of probabilities in this context is therefore a reflection of our lack of knowledge about the system, rather than something intrinsic to the system itself.

The probabilities in quantum theory are not like this at all; they are fundamental. It is not the case that we can only predict the probability of a particle being in one place or another because we are ignorant. We can't, *even in principle*, predict what the position of a particle will be. What we can predict, with absolute precision, is the probability that a particle will be found in a particular place if we look for it. More than that, we can predict with absolute precision how this probability changes with time. Born expressed this beautifully in 1926: 'The motion of particles follows probability laws but the probability itself propagates according to the law of causality.' This is exactly what Schrödinger's equation does: it is an equation that allows us to calculate exactly what the wavefunction will look like in the future, given what it looks like in the past. In that sense, it is analogous to Newton's laws. The difference is that, whilst Newton's laws allow us to calculate the position and speed of particles at any particular time in the future, quantum mechanics allows us to calculate only the probability that they will be found at a particular place.

This loss of predictive power was what bothered Einstein and many of his colleagues. With the benefit of over eighty years of hindsight and a great deal of hard work, the debate now seems somewhat redundant, and it is easy to dismiss it with the statement that Born, Heisenberg, Pauli, Dirac and others were correct and Einstein, Schrödinger and the old guard were wrong. But it was certainly possible back then to believe that quantum theory was incomplete in some way, and that the probabilities appear, just as in thermodynamics or coin tossing, because there is some information about the particles that we are missing. Today that idea gains little purchase – theoretical and experimental progress indicate that Nature really does use random numbers, and the loss of certainty in predicting the positions of particles is an intrinsic property of the physical world: probabilities are the best we can do.

4. *Everything That Can Happen Does Happen*

We've now set up a framework within which we can explore quantum theory in detail. The key ideas are very simple in their technical content, but tricky in the way they challenge us to confront our prejudices about the world. We have said that a particle is to be represented by lots of little clocks dotted around and that the length of the clock hand at a particular place (squared) represents the probability that the particle will be found at that place. The clocks are not the main point – they are a mathematical device we'll use to keep track of the odds on finding a particle somewhere. We also gave a rule for adding clocks together, which is necessary to describe the phenomenon of interference. We now need to tie up the final loose end, and look for the rule that tells us how the clocks change from one moment to the next. This rule will be the replacement of Newton's first law, in the sense that it will allow us to predict what a particle will do if we leave it alone. Let's begin at the beginning and imagine placing a single particle at a point.

Figure 4.1. The single clock representing a particle that is definitely located at a particular point in space.

We know how to represent a particle at a point, and this is shown in Figure 4.1. There will be a single clock at that point, with a hand of length 1 (because 1 squared is 1 and that means the probability to find the particle there is equal to 1, i.e. 100 per cent). Let's suppose that the clock reads 12 o'clock, although this choice is completely arbitrary. As far as the probability is concerned, the clock hand can point in any direction, but we have to choose something to start with, so 12 o'clock will do. The question we want to answer is the following: what is the chance that the particle will be located somewhere else at a later time? In other words, how many clocks do we have to draw, and where do we have to place them, at the next moment? To Isaac Newton, this would have been a very dull question; if we place a particle somewhere and do nothing to it, then it's not going to go anywhere. But Nature says, quite categorically, that this is simply wrong. In fact, Newton could not be more wrong.

Here is the correct answer: the particle *can be anywhere else in the Universe at the later time*. That means we have to draw an infinite number of clocks, one at every conceivable point in space. That sentence is worth reading lots of times. Probably we need to say more.

Allowing the particle to be anywhere at all is equivalent to assuming nothing about the motion of the particle. This is the most unbiased thing we can do, and that does have a certain ascetic appeal to it,[1] although admittedly it does seem to violate the laws of common sense, and perhaps the laws of physics as well.

A clock is a representation of something definite – the likelihood that a particle will be found at the position of the clock. If we know that a particle is at one particular place at a particular time, we represent it by a single clock at that point. The proposal is that if we start with a particle sitting at a definite position at time zero, then at 'time zero plus a little bit' we should draw a vast, indeed infinite, array of new clocks, filling the entire Universe. This admits the possibility that the particle hops off to *anywhere and everywhere* else in an instant. Our particle will simultaneously be both a nanometre away and also a billion light years away in the heart of a star in a distant galaxy. This sounds, to use our native northern vernacular, daft. But let's be very clear: the theory must be capable of explaining the double-slit experiment and, just as a wave spreads out if we dip a toe into still water, so an electron initially located somewhere

1. Or aesthetic appeal, depending on your point of view.

must spread out as time passes. What we need to establish is exactly how it spreads.

Unlike a water wave, we are proposing that the electron wave spreads out to fill the Universe in an instant. Technically speaking, we'd say that the rule for particle propagation is different from the rule for water wave propagation, although both propagate according to a 'wave equation'. The equation for water waves is different from the equation for particle waves (which is the famous Schrödinger equation we mentioned in the last chapter), but both encode wavy physics. The differences are in the details of how things propagate from place to place. Incidentally, if you know a little about Einstein's theory of relativity you might be getting nervous when we speak of a particle hopping across the Universe in an instant, because that would seem to correspond to something travelling faster than the speed of light. Actually, the idea that a particle can be here and, an instant later, somewhere else very far away is not in itself in contradiction with Einstein's theories, because the real statement is that *information* cannot travel faster than the speed of light, and it turns out that quantum theory remains constrained by that. As we shall learn, the dynamics corresponding to a particle leaping across the Universe are the very opposite of information transfer, because we cannot know where the particle will leap to beforehand. It seems we are building a theory on complete anarchy, and you might naturally be concerned that Nature surely cannot behave like this. But, as we shall see as the book unfolds, the order we see in the everyday world really does emerge out of this fantastically absurd behaviour.

If you are having trouble swallowing this anarchic proposal – that we have to fill the entire Universe with little clocks in order to describe the behaviour of a single subatomic particle from one moment to the next – then you are in good company. Lifting the veil on quantum theory and attempting to interpret its inner workings is baffling to everyone. Niels Bohr famously wrote that 'Those who are not shocked when they first come across quantum mechanics cannot possibly have understood it', and Richard Feynman introduced volume III of *The Feynman Lectures on Physics* with the words: 'I think I can safely say that nobody understands quantum mechanics.' Fortunately, following the rules is far simpler than trying to visualize what they actually mean. The ability to follow through the consequences of a particular set of assumptions carefully, without getting too hung up on the philosophical implications, is one of the most important skills a physicist learns. This is absolutely in the spirit of Heisenberg: let us set out our initial assumptions and compute their consequences. If we arrive at a set of predictions that agree with observations of the world around us, then we should accept the theory as good.

Many problems are far too difficult to solve in a single mental leap, and deep understanding rarely emerges in 'eureka' moments. The trick is to make sure that you understand each little step and after a sufficient number of steps the bigger picture should emerge. Either that or we realize we have been barking up the wrong tree and have to start over from scratch. The little steps we've outlined so far are not difficult in themselves, but the idea that we have decided to take a single clock and turn it into an infinity of clocks is certainly a tricky concept, especially if you try to imagine drawing them all. Eternity is a very long time, to paraphrase Woody Allen, especially near the end. Our advice is not to panic or give up and, in any case, the infinity bit is a detail. Our next task is to establish the rule that tells us what all those clocks should actually look like at some time after we laid down the particle.

The rule we are after is the essential rule of quantum theory, although we will need to add a second rule when we come to consider the possibility that the Universe contains more than just one particle. But first things first: for now, let's focus on a single particle alone in the Universe – no one can accuse us of rushing into things. At one instant in time, we'll suppose we know exactly where it is, and it's therefore represented by a single, solitary clock. Our specific task is to identify the rule that will tell us what each and every one of the new clocks, scattered around the Universe, should look like at any time in the future.

We'll first state the rule without any justification. We will come back to discuss just why the rule looks like it does in a few paragraphs, but for now we should treat it as one of the rules in a game. Here's the rule: at a time t in the future, a clock a distance x from the original clock has its hand wound in an anti-clockwise direction by an amount proportional to x^2; the amount of winding is also proportional to the mass of the particle m and inversely proportional to the time t. In symbols, this means we are to wind the clock hand anti-clockwise by an amount proportional to mx^2/t. In words, it means that there is more winding for a more massive particle, more winding the further away the clock is from the original, and less winding for a bigger step forward in time. This is an algorithm – a recipe if you like – that tells us exactly what to do to work out what a given arrangement of clocks will look like at some point in the future. At every point in the universe, we draw a new clock with its hand wound around by an amount given by our rule. This accounts for our assertion that the particle can, and indeed does, hop from its initial position to each and every other point in the Universe, spawning new clocks in the process.

To simplify matters we have imagined just one initial clock, but of course at some instant in time there might already be many clocks, representing the fact that the particle is not at some definite location. How are we to figure out what to do with a whole cluster of clocks? The answer is that we are to do what we did for one clock, and repeat that for each and every one of the clocks in the cluster. Figure 4.2 illustrates this idea. The initial set of clocks are represented by the little circles, and the arrows indicate that the particle hops from the site of every initial clock to the point X, 'depositing' a new clock in the process. Of course, this delivers one new clock to X for every initial clock, and we must add all these clocks together in order to construct the final, definitive clock at X. The size of this final clock's hand gives us the chance of finding the particle at X at the later time.

It is not so strange that we should be adding clocks together when several arrive at the same point. Each clock corresponds to a different way that the particle could have reached X. This addition of the clocks is understandable if we think back to the double-slit experiment; we are simply trying to rephrase the wave description in terms of clocks. We can imagine two initial clocks, one at each slit. Each of these two clocks will deliver a clock to a particular point on the screen at some later time, and we must add these two clocks together in order to obtain the interference pattern.[2] In summary therefore, the rule to calculate what the clock looks like at any point is to transport all the initial clocks to that point, one by one, and then add them together using the addition rule we encountered in the previous chapter.

Since we developed this language in order to describe the propa-

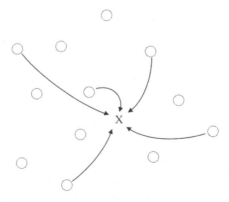

Figure 4.2. Clock hopping. The open circles indicate the locations of the particle at some instant in time; we are to associate a clock with each point. To compute the probability to find the particle at X we are to allow the particle to hop there from all of the original locations. A few such hops are indicated by the arrows. The shape of the lines does not have any meaning and it certainly does not mean that the particle travels along some trajectory from the site of a clock to X.

2. If you are having trouble with that last sentence try replacing the word 'clock' with 'wave'.

gation of waves, we can also think about more familiar waves in these terms. The whole idea, in fact, goes back a long way. Dutch physicist Christiaan Huygens famously described the propagation of light waves like this as far back as 1690. He did not speak about imaginary clocks, but rather he emphasized that we should regard each point on a light wave as a source of secondary waves (just as each clock spawns many secondary clocks). These secondary waves then combine to produce a new resultant wave. The process repeats itself so that each point in the new wave also acts as a source of further waves, which again combine, and in this way a wave advances.

We can now return to something that may quite legitimately have been bothering you. Why on earth did we choose the quantity mx^2/t to determine the amount of winding of the clock hand? This quantity has a name: it is known as the *action*, and it has a long and venerable history in physics. Nobody really understands why Nature makes use of it in such a fundamental way, which means that nobody can really explain why those clocks get wound round by the amount they do. Which somewhat begs the question: how did anyone realize it was so important in the first place? The action was first introduced by the German philosopher and mathematician Gottfried Leibniz in an unpublished work written in 1669, although he did not find a way to use it to make calculations. It was reintroduced by the French scientist Pierre-Louis Moreau de Maupertuis in 1744, and subsequently used to formulate a new and powerful principle of Nature by his friend, the mathematician Leonard Euler. Imagine a ball flying through the air. Euler found that the ball travels on a path such that the action computed between any two points on the path is always the smallest that it can be. For the case of a ball, the action is related to the *difference* between the kinetic and potential energies of the ball.[3] This is known as 'the principle of least action',

---

3. The kinetic energy is equal to $mv^2/2$ and the potential energy is mgh when the ball is a height h above the ground. g is the rate at which all objects accelerate in

and it can be used to provide an alternative to Newton's laws of motion. At first sight it's a rather odd principle, because in order to fly in a way that minimizes the action, the ball would seem to have to know where it is going before it gets there. How else could it fly through the air such that, when everything is done, the quantity called the action is minimized? Phrased in this way, the principle of least action sounds teleological – that is to say things appear to happen in order to achieve a pre-specified outcome. Teleological ideas generally have a rather bad reputation in science, and it's easy to see why. In biology, a teleological explanation for the emergence of complex creatures would be tantamount to an argument for the existence of a designer, whereas Darwin's theory of evolution by natural selection provides a simpler explanation that fits the available data beautifully. There is no teleological component to Darwin's theory – random mutations produce variations in organisms, and external pressures from the environment and other living things determine which of these variations are passed on to the next generation. This process alone can account for the complexity we see in life on Earth today. In other words, there is no need for a grand plan and no gradual assent of life towards some sort of perfection. Instead, the evolution of life is a random walk, generated by the imperfect copying of genes in a constantly shifting external environment. The Nobel-Prize-winning French biologist Jacques Monod went so far as to define a cornerstone of modern biology as 'the systematic or axiomatic denial that scientific knowledge can be obtained on the basis of theories that involve, explicitly or not, a teleological principle'.

As far as physics is concerned, there is no debate as to whether or not the least action principle actually works, for it allows calculations to be performed that correctly describe Nature and it is a cornerstone of physics. It can be argued that the least action principle is not teleological at all, but the debate is in any case neutralized

---

the vicinity of the Earth. The action is their difference integrated between the times associated with the two points on the path.

once we have a grasp of Feynman's approach to quantum mechanics. The ball flying through the air 'knows' which path to choose because it actually, secretly, explores every possible path.

How was it discovered that the rule for winding the clocks should have anything to do with this quantity called the action? From a historical perspective, Dirac was the first to search for a formulation of quantum theory that involved the action, but rather eccentrically he chose to publish his research in a Soviet journal, to show his support for Soviet science. The paper, entitled 'The Lagrangian in Quantum Mechanics', was published in 1933 and languished in obscurity for many years. In the spring of 1941, the young Richard Feynman had been thinking about how to develop a new approach to quantum theory using the Lagrangian formulation of classical mechanics (which is the formulation derived from the principle of least action). He met Herbert Jehle, a visiting physicist from Europe, at a beer party in Princeton one evening, and, as physicists tend to do when they've had a few drinks, they began discussing research ideas. Jehle remembered Dirac's obscure paper, and the following day they found it in the Princeton Library. Feynman immediately started calculating using Dirac's formalism and, in the course of an afternoon with Jehle looking on, he found that he could derive the Schrödinger equation from an action principle. This was a major step forward, although Feynman initially assumed that Dirac must have done the same because it was such an easy thing to do; easy, that is, if you are Richard Feynman. Feynman eventually asked Dirac whether he'd known that, with a few additional mathematical steps, his 1933 paper could be used in this way. Feynman later recalled that Dirac, lying on the grass in Princeton after giving a rather lacklustre lecture, simply replied, 'No, I didn't know. That's interesting.' Dirac was one of the greatest physicists of all time, but a man of few words. Eugene Wigner, himself one of the greats, commented that 'Feynman is a second Dirac, only this time human.'

To recap: we have stated a rule that allows us to write down the whole array of clocks representing the state of a particle at some instant in time. It's a bit of a strange rule – fill the Universe with an infinite number of clocks, all turned relative to each other by an amount that depends on a rather odd but historically important quantity called the action. If two or more clocks land at the same point, add them up. The rule is built on the premise that we must allow a particle the freedom to jump from any particular place in the Universe to absolutely anywhere else in an infinitesimally small moment. We said at the outset that these outlandish ideas must ultimately be tested against Nature to see whether anything sensible emerges. To make a start on that, let's see how something very concrete, one of the cornerstones of quantum theory, emerges from this apparent anarchy: Heisenberg's Uncertainty Principle.

Heisenberg's Uncertainty Principle

Heisenberg's Uncertainty Principle is one of the most misunderstood parts of quantum theory, a doorway through which all sorts of charlatans and purveyors of tripe[4] can force their philosophical musings. He presented it in 1927 in a paper entitled 'Über den anschaulichen Inhalt der quantentheoretischen Kinematik und Mechanik', which is very difficult to translate into English. The difficult word is *anschaulich*, which means something like 'physical' or 'intuitive'. Heisenberg seems to have been motivated by his intense annoyance that Schrödinger's more intuitive version of quantum theory was more widely accepted than his own, even though both formalisms led to the same results. In the spring of 1926, Schrödinger was convinced that his equation for the wavefunction provided a physical picture of what was going on inside atoms. He thought that his wavefunction was a thing you could visualize, and was related to the distribution of electric charge inside the atom. This turned out to be incorrect, but at least it made physicists feel good during the

4. Wikipedia describes 'tripe' as 'a type of edible offal from the stomachs of various farm animals', but it is colloquially used to mean 'nonsense'. Either definition is appropriate here.

first six months of 1926: until Born introduced his probabilistic interpretation.

Heisenberg, on the other hand, had built his theory around abstract mathematics, which predicted the outcomes of experiments extremely successfully but was not amenable to a clear physical interpretation. Heisenberg expressed his irritation to Pauli in a letter on 8 June 1926, just weeks before Born threw his metaphorical spanner into Schrödinger's intuitive approach. 'The more I think about the physical part of Schrödinger's theory, the more disgusting I find it. What Schrödinger writes about the *Anschaulichkeit* of his theory . . . I consider *Mist.*' The translation of the German word *mist* is 'rubbish' or 'bullshit' . . . or 'tripe'.

What Heisenberg decided to do was to explore what an 'intuitive picture', or *Anschaulichkeit*, of a physical theory should mean. What, he asked himself, does quantum theory have to say about the familiar properties of particles such as position? In the spirit of his original theory, he proposed that a particle's position is a meaningful thing to talk about only if you also specify how you measure it. So you can't ask where an electron actually is inside a hydrogen atom without describing exactly how you'd go about finding out that information. This might sound like semantics, but it most definitely is not. Heisenberg appreciated that the very act of measuring something introduces a disturbance, and that as a result there is a limit on how well we can 'know' an electron. Specifically, in his original paper, Heisenberg was able to estimate what the relationship is between how accurately we can simultaneously measure the position and the momentum of a particle. In his famous Uncertainty Principle, he stated that if Δx is the uncertainty in our knowledge of the position of a particle (the Greek letter Δ is pronounced 'delta', so Δx is pronounced 'delta x') and Δp is the corresponding uncertainty in the momentum, then

$$\Delta x \Delta p \sim h$$

where h is Planck's constant and the '$\sim$' symbol means 'is similar in size to'. In words, the product of the uncertainty in the position of a particle and the uncertainty in its momentum will be roughly equal to Planck's constant. This means that the more accurately we identify the location of a particle, the less well we can know its momentum, and vice versa. Heisenberg came to this conclusion by contemplating the scattering of photons off electrons. The photons are the means by which we 'see' the electron, just as we see everyday objects by scattering photons off them and collecting them in our eyes. Ordinarily, the light that bounces off an object disturbs the object imperceptibly, but that is not to deny our fundamental inability to absolutely isolate the act of measurement from the thing one is measuring. One might worry that it could be possible to beat the limitations of the Uncertainty Principle by devising a suitably ingenious experiment. We are about to show that this is not the case and the Uncertainty Principle is absolutely fundamental, because we are going to derive it using only our theory of clocks.

Deriving Heisenberg's Uncertainty Principle from the Theory of Clocks

Rather than starting with a particle at a single point, let us instead think about a situation where we know roughly where the particle is, but we don't know exactly where it is. If we know that a particle is somewhere in a small region of space then we should represent it by a cluster of clocks filling that region. At each point within the region there will be a clock, and that clock will represent the probability that the particle will be found at that point. If we square up the lengths of all the clock hands at every point and add them together, we will get 1, i.e. the probability to find the particle *somewhere* in the region is 100 per cent.

In a moment we are going to use our quantum rules to perform a serious calculation, but first we should come clean and say that we have failed to mention an important addendum to the clock-winding rule. We didn't want to introduce it earlier because it is a technical detail, but we won't get the correct answers when it comes to calculating actual probabilities if we ignore it. It relates to what we said at the end of the previous paragraph.

If we begin with a single clock, then the hand must be of length 1, because the particle must be found at the location of the clock with a probability of 100 per cent. Our quantum rule then says that, in order to describe the particle at some later time, we should transport this clock to all points in the Universe, corresponding to the particle leaping from its initial location. Clearly we cannot leave all of the clock hands with a length of 1, because then our probability interpretation falls down. Imagine, for example, that the particle is described by four clocks, corresponding to its being at four different locations. If each one has a size of 1 then the probability that the particle is located at any one of the four positions would be 400 per cent and this is obviously nonsense. To fix this problem we must shrink the clocks in addition to winding them anti-clockwise. This 'shrink rule' states that after all of the new clocks have been spawned, every clock should be shrunk by the square root of the total number of clocks.[5] For four clocks, that would mean that each hand must be shrunk by $\sqrt{4}$, which means that each of the four final clocks will have a hand of length $\frac{1}{2}$. There is then a $(\frac{1}{2})^2 = 25$ per cent chance that the particle will be found at the site of any one of the four clocks. In this simple way we can ensure that the probability that the particle is found somewhere will always total 100 per cent. Of course, there may be an infinite number of possible locations, in which case the clocks would have zero size, which may sound alarming, but the maths can handle it. For our purposes, we shall always imagine that there are a finite number of clocks, and in any case we will never actually need to know how much a clock shrinks.

Let's get back to thinking about a Universe containing a single particle whose location is not precisely known. You can treat the

---

5. Shrinking all clocks by the same amount is strictly only true provided that we are ignoring the effects of Einstein's Special Theory of Relativity. Otherwise, some of the clocks get shrunk more than others. We shan't need to worry about this.

next section as a little mathematical puzzle – it may be tricky to follow the first time through, and it may be worth rereading, but if you are able to follow what is going on then you'll understand how the Uncertainty Principle emerges. For simplicity, we've assumed that the particle moves in one dimension, which means it is located somewhere on a line. The more realistic three-dimensional case is not fundamentally different – it's just harder to draw. In Figure 4.3 we've sketched this situation, representing the particle by a line of three clocks. We should imagine that there are many more than this – one at every possible point that the particle could be – but this would be very hard to draw. Clock 3 sits at the left side of the initial clock cluster and clock 1 is at the right side. To reiterate, this represents a situation in which we know that the particle starts out somewhere between clocks 1 and 3. Newton would say that the particle stays between clocks 1 and 3 if we do nothing to it, but what does the quantum rule say? This is where the fun starts – we are going to play with the clock rules to answer this question.

Figure 4.3. A line of three clocks all reading the same time: this describes a particle initially located in the region of the clocks. We are interested in figuring out what the chances are of finding the particle at the point X at some later time.

Let's allow time to tick forward and work out what happens to this line of clocks. We'll start off by thinking about one particular point a large distance away from the initial cluster, marked X in the figure. We'll be more quantitative about what a 'large distance' means later on, but for now it simply means that we need to do a lot of clock winding.

Applying the rules of the game, we should take each clock in the initial cluster and transport it to point X, winding the hand around and shrinking it accordingly. Physically, this corresponds to the particle hopping from that point in the cluster to point X. There will be many clocks arriving at X, one from each initial clock in the line, and we should add them all up. When all this is done, the square of the length of the resulting clock hand at X will give the probability that we will find the particle at X.

Now let's see how this all pans out and put some numbers in. Let's say that the point X is a distance of '10 units' away from clock 1, and that the initial cluster is '0.2 units' wide. Answering the obvious question: 'How far is 10 units?' is where Planck's constant enters our story, but for now we shall deftly side-step that issue and simply specify that 1 unit of distance corresponds to 1 complete (twelve-hour) wind of the clock. This means that the point X is approximately $10^2 = 100$ complete windings away from the initial cluster (remember the winding rule). We shall also assume that the clocks in the initial cluster started out of equal size, and that they all point to 12 o'clock. Assuming they are of equal size is simply the assumption that the particle is equally likely to be anywhere in between points 1 and 3 in the figure and the significance of them all reading the same time will emerge in due course.

To transport a clock from point 1 to point X, we have to rotate the clock hand anti-clockwise 100 complete times, as per our rule. Now let's move across to point 3, which is a further 0.2 units away, and transport that clock to X. This clock has to travel 10.2 units, so we have to wind its hand back a little more than before, i.e. by 10.2^2, which is very close to 104, complete winds.

We now have two clocks landing at X, corresponding to the particle hopping from 1 to X and from 3 to X, and we must add them together in order to start the task of computing the final clock. Because they both got wound around by very close to a whole number of winds, they will both end up pointing roughly to 12 o'clock, and they will add up to form a clock with a bigger hand also pointing to 12 o'clock. Notice that it is only the final direction of the clock hands that matters. We do not need to keep track of how often they wind around. So far so good, but we haven't finished because there are many other little clocks in between the right- and left-hand edges of the cluster.

And so our attention now turns to the clock midway between the two edges, i.e. at point 2. That clock is 10.1 units away from X, which means that we have to wind it 10.1^2 times. This is very close to 102 complete rotations – again a whole number of winds. We need to add this clock to the others at X and, as before, this will make the hand at X even longer. Continuing, there is also a point midway between points 1 and 2 and the clock hopping from there will get 101 complete rotations, which will add to the size of the final hand again. But here is the important point. If we now go midway again between these two points, we get to a clock that will be wound 100.5 rotations when it reaches X. This corresponds to a clock with a hand pointing to 6 o'clock, and when we add this clock we will *reduce* the length of the clock hand at X. A little thought should convince you that, although the points labelled 1, 2 and 3 each produce clocks at X reading 12 o'clock, and although the points midway between 1, 2 and 3 also produce clocks that read 12 o'clock, the points that are $\frac{1}{4}$ and $\frac{3}{4}$ of the way between points 1 and 3 and points 2 and 3 each generate clocks that point to 6 o'clock. In total that is five clocks pointing up and four clocks pointing down. When we add all these clocks together, we'll get a resultant clock at X that has a tiny hand because nearly all of the clocks will cancel each other out.

This 'cancellation of clocks' obviously extends to the realistic case where we consider every possible point lying in the region between points 1 and 3. For example, the point that lies $\frac{1}{8}$ of the way along from point 1 contributes a clock reading 9 o'clock, whilst the point lying $\frac{3}{8}$ of the way reads 3 o'clock – again the two cancel each other out. The net effect is that the clocks corresponding to all of the ways that the particle could have travelled from somewhere in the cluster to point X cancel each other out. This cancellation is illustrated on the far right of the figure. The arrows indicate the clock hands arriving at X from various points in the initial cluster.

The net effect of adding all these arrows together is that they all cancel each other out. This is the crucial 'take home' message.

To reiterate, we have just shown that, provided the original cluster of clocks is large enough and that point X is far enough away, then for every clock that arrives at X pointing to 12 o'clock, there will be another that arrives pointing to 6 o'clock to cancel it out. For every clock that arrives pointing to 3 o'clock, there will be another that arrives pointing to 9 o'clock to cancel it out, and so on. This wholesale cancellation means that there is effectively no chance at all of finding the particle at X. This really is very encouraging and interesting, because it looks rather like a description of a particle that isn't moving. Although we started out with the ridiculous-sounding proposal that a particle can go from being at a single point in space to anywhere else in the Universe a short time later, we have now discovered that this is not the case if we start out with a cluster of clocks. For a cluster, because of the way all the clocks interfere with each other, the particle has effectively no chance of being far away from its initial position. This conclusion has come about as a result of an 'orgy of quantum interference', in the words of Oxford professor James Binney.

For the orgy of quantum interference and corresponding cancellation of clocks to happen, point X needs to be far enough away from the initial cluster so that the clocks can rotate around many times. Why? Because if point X is too close then the clock hands won't necessarily have the chance to go around at least once, which means they will *not* cancel each other out so effectively. Imagine, for example, that the distance from the clock at point 1 to point X is 0.3 instead of 10. Now the clock at the front of the cluster gets a smaller wind than before, corresponding to $0.3^2 = 0.09$ of a turn, which means it is pointing just past 1 o'clock. Likewise, the clock from point 3, at the back of the cluster, now gets wound by $0.5^2 = 0.25$ of a turn, which means it reads 3 o'clock. Consequently, all of the clocks arriving at X point somewhere between 1 o'clock and 3 o'clock, which means they do not cancel each other out but instead add up to one big clock pointing to approximately 2 o'clock.

All of this amounts to saying that there will be a reasonable chance of finding the particle at points close to, but outside of, the original cluster. By 'close to', we mean that there isn't sufficient winding to get the clock hands around at least once. This is starting to have a whiff of the Uncertainty Principle about it, but it is still a little vague, so let's explore exactly what we mean by a 'large enough' initial cluster and a point 'far enough away'.

Our initial ansatz, following Dirac and Feynman, was that the amount the hands wind around when a particle of mass m hops a distance x in a time t is proportional to the action, i.e. the amount of winding is proportional to mx^2/t. Saying it is 'proportional to' isn't good enough if we want to calculate real numbers. We need to know precisely what the amount of winding is equal to. In Chapter 2 we discussed Newton's law of gravitation, and in order to make quantitative predictions we introduced Newton's gravitational constant, which determines the strength of the gravitational force. With the addition of Newton's constant, numbers can be put into the equation and real things can be calculated, such as the orbital period of the Moon or the path taken by the Voyager 2 spacecraft on its journey across the solar system. We now need something similar for quantum mechanics – a constant of Nature that 'sets the scale' and allows us to take the action and produce a precise statement about the amount by which we should wind clocks as we move them a specified distance away from their initial position in a particular time. That constant is Planck's constant.

A Brief History of Planck's Constant

In a flight of imaginative genius during the evening of 7 October 1900, Max Planck managed to explain the way that hot objects radiate energy. Throughout the second half of the nineteenth century, the exact relationship between the distribution of the wavelengths of light emitted by hot objects and their temperature was one of the great puzzles in physics. Every hot object emits light and, as the temperature is increased, the character of the light changes. We are familiar with light in the visible region, corresponding to the colours of the rainbow, but light can also occur with wavelengths that are either too long or too short to be seen by the human eye. Light with a longer wavelength than red light is called 'infra-red' and it can be seen using night-vision goggles. Still longer wavelengths correspond to radio waves. Likewise, light with a wavelength just shorter than blue is called ultra-violet, and the shortest wavelength light is generically referred to as 'gamma radiation'. An unlit lump of coal at room temperature will emit light in the infra-red part of the spectrum. But if we throw it on to a burning fire, it will begin to glow red. This is because, as the temperature of the coal rises, the average wavelength of the radiation it emits decreases, eventually entering the range that our eyes can see. The rule is that the hotter the object, the shorter the wavelength of the light it emits. As the precision of the experimental measurements improved in the nineteenth century, it became clear that nobody had the correct mathematical formula to describe this observation. This problem is often termed the 'black body problem', because physicists refer to idealized objects that perfectly absorb and then re-emit radiation as 'black bodies'. The problem was a serious one, because it revealed an inability to understand the character of light emitted by anything and everything.

Planck had been thinking hard about this and related matters in the fields of thermodynamics and electromagnetism for many years before he was appointed Professor of Theoretical Physics in Berlin. The post had been offered to both Boltzmann and Hertz before Planck was approached, but both declined. This proved to be fortuitous, because Berlin was the centre of the experimental investigations into black body radiation, and Planck's immersion at the heart of the experimental work proved key to his subsequent theoretical tour de force. Physicists often work best when they are able to have wide-ranging and unplanned conversations with colleagues.

We know the date and time of Planck's revelation so well because he and his family had spent the afternoon of Sunday 7 October 1900 with his colleague Heinrich Rubens. Over lunch, they discussed the failure of the theoretical models of the day to explain the details of black body radiation. By the evening, Planck had scribbled a formula on to a postcard and sent it to Rubens. It turned out to be the correct formula, but it was very strange indeed. Planck later described it as 'an act of desperation', having tried everything else he could think of. It is genuinely unclear how Planck came up with his formula. In his superb biography of Albert Einstein, *Subtle is the Lord ...*, Abraham Pais writes: 'His reasoning was mad, but his madness has that divine quality that only the greatest transitional figures can bring to science.' Planck's proposal was both inexplicable and revolutionary. He found that he could explain the black body spectrum, but only if he assumed that the energy of the emitted light was made up of a large number of smaller 'packets' of energy. In other words the total energy is quantized in units of a new fundamental constant of Nature, which Planck called 'the quantum of action'. Today, we call it Planck's constant.

What Planck's formula actually implies, although he didn't appreciate it at the time, is that light is *always* emitted and absorbed in packets, or quanta. In modern notation, those packets have energy $E = hc/\lambda$, where λ is the wavelength of the light (pronounced 'lambda'), c is the speed of light and h is Planck's constant. The role of Planck's constant in this equation is as the conversion factor between the wavelength of light and the energy of its associated quantum. The realization that the quantization of the energy of emitted light, as identified by Planck, arises because the light itself is made up of particles was proposed, tentatively at first, by Albert Einstein. He made the proposition during his great burst of creativity in 1905 – the annus mirabilis which also produced the Special Theory of Relativity and the most famous equation in scientific history, $E = mc^2$. Einstein received the 1921 Nobel Prize for physics (which due to a rather arcane piece of Nobelian bureaucracy he received in 1922) for this work on the photoelectric effect, and not for his better-known theories of relativity. Einstein proposed that light can be regarded as a stream of particles (he did not at that time use the word 'photons') and he correctly recognized that the energy of each photon is inversely proportional to its wavelength. This conjecture by Einstein is the origin of one of the most famous paradoxes in quantum theory – that particles behave as waves, and vice versa.

Planck removed the first bricks from the foundations of Maxwell's picture of light by showing that the energy of the light emitted from a hot object can only be described if it is emitted in quanta. It was Einstein who pulled out the bricks that brought down the whole edifice of classical physics. His interpretation of the photoelectric effect demanded not only that light is emitted in little packets, but that it also interacts with matter in the form of localized packets. In other words, light really does behave as a stream of particles.

The idea that light is made from particles – that is to say that 'the electromagnetic field is quantized' – was deeply controversial and not accepted for decades after Einstein first proposed it. The reluctance of Einstein's peers to embrace the idea of the photon can be seen in the proposal, co-written by Planck himself, for Einstein's membership of the prestigious Prussian Academy in 1913, a full eight years after Einstein's introduction of the photon:

> In sum, one can say that there is hardly one among the great problems in which modern physics is so rich to which Einstein has not made a remarkable contribution. That he may sometimes have missed the target in his speculations, as, for example, in his hypothesis of light quanta, cannot really be held too much against him, for it is not possible to introduce really new ideas even in the most exact sciences without sometimes taking a risk.

In other words, nobody really believed that photons were real. The widely held belief was that Planck was on safe ground because his proposal was more to do with the properties of matter – the little oscillators that emitted the light – rather than the light itself. It was simply too strange to believe that Maxwell's beautiful wave equations needed replacing with a theory of particles.

We mention this history partly to reassure you of the genuine difficulties that must be faced in accepting quantum theory. It is impossible to visualize a thing, such as an electron or a photon, that behaves a little bit like a particle, a little bit like a wave, and a little bit like neither. Einstein remained concerned about these issues for the rest of his life. In 1951, just four years before his death, he wrote: 'All these fifty years of pondering have not brought me any closer to answering the question, what are light quanta?'

Sixty years later, what is unarguable is that the theory we are in the process of developing using our arrays of little clocks describes, with unerring precision, the results of every experiment that has ever been devised to test it.

Back to Heisenberg's Uncertainty Principle

This, then, is the history behind the introduction of Planck's constant. But for our purposes, the most important thing to notice is that Planck's constant is a unit of 'action', which is to say that it is the same type of quantity as the thing which tells us how far to wind the clocks. Its modern value is $6.6260695729 \times 10^{-34}$ kg m^2/s, which is very tiny by everyday standards. This will turn out to be the reason why we don't notice its all-pervasive effects in everyday life.

Recall that we wrote of the action corresponding to a particle hopping from one place to another as the mass of the particle multiplied by the distance of the hop squared divided by the time interval over which the hop occurs. This is measured in kg m^2/s, as is Planck's constant, and so if we simply divide the action by Planck's constant, we'll cancel all the units out and end up with a pure number. According to Feynman, this pure number is the amount we should wind the clock associated with a particle hopping from one place to another. For example, if the number is 1, that means 1 full wind and if it's $\frac{1}{2}$, it means $\frac{1}{2}$ a wind, and so on. In symbols, the precise amount by which we should turn the clock hand to account for the possibility that a particle hops a distance x in a time t is $mx^2/(2ht)$.

Notice that a factor $\frac{1}{2}$ has appeared in the formula. You can either take that as being what is needed to agree with experiment or you can note that this arises from the definition of the action.[6] Either is fine. Now that we know the value of Planck's constant, we can really quantify the amount of winding and address the point we deferred a little earlier. Namely, what does jumping a distance of '10' actually mean?

Let's see what our theory has to say about something small by everyday standards: a grain of sand. The theory of quantum mechanics we've developed suggests that if we place the grain down somewhere then at a later time it *could* be anywhere in the Universe. But this is obviously not what happens to real grains of sand. We have already glimpsed a way out of this potential problem because if there is sufficient interference between the clocks, corresponding to the sand grain hopping from a variety of initial locations, then they will all cancel out to leave the grain sitting still. The first question we need to answer is: how many times will the clocks get wound if we transport a particle with the mass of a grain of sand a distance of, say, 0.001 millimetres, in a time of one second? We wouldn't be able to see such a tiny distance with our eyes, but it is still quite large on the scale of atoms. You can do the calculation quite easily yourself by substituting the numbers into Feynman's winding rule.[7] The answer is something like a trillion complete winds of the clock. Imagine how much interference that

6. For a particle of mass m that hops a distance x in a time t, the action is $\frac{1}{2}m(x/t)^2 t$ if the particle travels in a straight line at constant speed. But this does not mean the quantum particle travels from place to place in straight lines. The clock-winding rule is obtained by associating a clock with each possible path the particle can take between two points and it is an accident that, after summing over all these paths, the result is equal to this simple result. For example, the clock-winding rule is not this simple if we include corrections to ensure consistency with Einstein's Theory of Special Relativity.
7. A sand grain typically has a mass around 1 microgram, which is a billionth of a kilogram.

allows for. The upshot is that the sand grain stays where it is and there is almost no probability that it will jump a discernible distance, even though we really have to consider the possibility that it secretly hopped everywhere in the Universe in order to reach that conclusion.

This is a very important result. If you had put the numbers in for yourself then you'd already have a feel for why this is the case; it's the smallness of Planck's constant. Written out in full, it has a value $0.000000000000000000000000000000000066260695729$ kg m²/s. Dividing pretty much any everyday number by that will result in a lot of clock winding and a lot of interference, with the result that the exotic journeys of our sand grain across the Universe all cancel each other out, and we perceive this voyager through infinite space as a boring little speck of dust sitting motionless on a beach.

Our particular interest of course is in those circumstances where clocks do not cancel each other out, and, as we have seen, this occurs if the clocks do not turn by more than a single wind. In that case, the orgy of interference will not happen. Let's see what this means quantitatively.

Figure 4.4. The same as Figure 4.3 except that we are now not committing to a specific value of the size of the clock cluster or the distance to the point X.

We are going to return to the clock cluster, which we've redrawn in Figure 4.4, but we'll be more abstract in our analysis this time instead of committing to definite numbers. We will suppose that the cluster has a size equal to Δx, and the distance of the closest point in the cluster to point X is x. In this case, the cluster size Δx refers to the uncertainty in our knowledge of the initial position of the particle; it started out somewhere in a region of size Δx. Starting with point 1, the point in the cluster closest to point X, we should wind the clock corresponding to a hop from this point to X by an amount

$$W_1 = \frac{mx^2}{2ht}$$

Now let's go to the farthest point, point 3. When we transport the clock from this point to X, it will be wound around by a greater amount, i.e.

$$W_3 = \frac{m(x + \Delta x)^2}{2ht}$$

We can now be precise and state the condition for the clocks propagated from all points in the cluster *not* to cancel out at X: there should be less than one full wind of difference between the clocks from points 1 and 3, i.e.

$$W_3 - W_1 < \text{one wind}$$

Writing this out in full, we have

$$\frac{m(x + \Delta x)^2}{2ht} - \frac{mx^2}{2ht} < 1$$

We're now going to consider the specific case for which the cluster size, Δx, is much smaller than the distance x. This means we are asking for the prospects that our particle will make a leap far outside of its initial domain. In this case, the condition for no clock cancellation, derived directly from the previous equation, is

$$\frac{mx\Delta x}{ht} < 1$$

If you know a little maths, you'll be able to get this by multiplying out the bracketed term and neglecting all the terms that involve $(\Delta x)^2$. This is a valid thing to do because we've said that Δx is very small compared to x, and a small quantity squared is a very small quantity.

This equation is the condition for there to be no cancellation

69

of the clocks at point X. We know that if the clocks don't cancel out at a particular point, then there is a good chance that we will find the particle there. So we have discovered that if the particle is initially located within a cluster of size Δx, then at a time t later there is a good chance to find it a long distance x away from the cluster if the above equation is satisfied. Furthermore, this distance increases with time, because we are dividing by the time t in our formula. In other words, as more time passes, the chances of finding the particle further away from its initial position increases. This is beginning to look suspiciously like a particle that is moving. Notice also that the chance of finding the particle a long way away also increases as Δx gets smaller – i.e. as the uncertainty in the initial position of the particle gets smaller. In other words, the more accurately we pin down the particle, the faster it moves away from its initial position. This now looks a lot like Heisenberg's Uncertainty Principle.

To make final contact, let us rearrange the equation a little bit. Notice that for a particle to make its way from anywhere in the cluster to point X in time t, it must leap a distance x. If you actually measured the particle at X then you would naturally conclude that the particle had travelled at a speed equal to x/t. Also, remember that the mass multiplied by the speed of a particle is its momentum, so the quantity mx/t is the measured momentum of the particle. We can now go ahead and simplify our equation some more, and write

$$\frac{p\Delta x}{h} < 1$$

where p is the momentum. This equation can be rearranged to read

$$p\Delta x < h$$

and this really is important enough to merit more discussion, because it looks very much like Heisenberg's Uncertainty Principle.

This is the end of the maths for the time being, and if you haven't followed it too carefully you should be able to pick the thread up from here.

If we start out with a particle localized within a blob of size Δx,

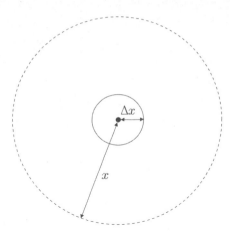

Figure 4.5. A small cluster grows with time, corresponding to a particle that is initially localized becoming delocalized as time advances.

we have just discovered that, after some time has passed, it could be found anywhere in a larger blob of size x. The situation is illustrated in Figure 4.5. To be precise, this means that if we had looked for the particle initially, then the chances are that we would have found it somewhere inside the inner blob. If we didn't measure it but instead waited a while, then there would be a good chance of finding it later on anywhere within the larger blob. This means that the particle could have moved from a position within the small initial blob to a position within the larger one. It doesn't have to have moved, and there is still a probability that it will be within the smaller region Δx. But it is quite possible that a measurement will reveal that the particle has moved as far out as the edge of the bigger blob.[8] If this extreme case were realized in a measurement then we would conclude that the particle is moving with a momentum given by the equation we just derived (and if you have not followed the maths then you will just have to take this on trust), i.e. $p = h/\Delta x$.

8. There is a chance that the particle travels even farther than the 'extreme' case marked out by the large blob in the figure but, as we have shown, the clocks tend to cancel out for such scenarios.

Now, we could start from the beginning again and set everything up exactly as before, so that the particle is once again initially located in the smaller blob of size Δx. Upon measuring the particle, we would probably find it somewhere else inside the larger blob, other than the extreme edge, and would therefore conclude that its momentum is smaller than the extreme value.

If we imagine repeating this experiment again and again, measuring the momentum of a particle that starts out inside a small cluster of size Δx, then we will typically measure a range of values of p anywhere between zero and the extreme value $h/\Delta x$. Saying that 'if you do this experiment many times then I predict you will measure the momentum to be somewhere between zero and $h/\Delta x$' means that 'the momentum of the particle is uncertain by an amount $h/\Delta x$'. Just as for the case of the uncertainty in position, physicists assign the symbol Δp to this uncertainty, and write $\Delta p \Delta x \sim h$. The '$\sim$' sign indicates that the product of the uncertainties in position and momentum is roughly equal to Planck's constant – it might be a little bigger or it might be a little smaller. With a little more care in the mathematics we could get this equation exactly right. The result would depend upon the details of the initial clock cluster, but it is not worth the extra effort to spend time doing that because what we have done is sufficient to capture the key ideas.

The statement that the uncertainty in a particle's position multiplied by the uncertainty in its momentum is (approximately) equal to Planck's constant is perhaps the most familiar form of Heisenberg's Uncertainty Principle. It is telling us that, starting from the knowledge that the particle is located within some region at some initial time, a measurement of the particle's position at some time later will reveal that the particle is moving with a momentum whose value cannot be predicted more accurately than 'somewhere between zero and $h/\Delta x$'. In other words, if we start out by confining a particle to be in a smaller and smaller region, then it has a tendency to want to jump further and further away from that region. This is so important, it is worth restating a third time: the more precisely you know the position of a particle at some instant, the less well you know how fast it is moving and therefore where it will be sometime later.

This is exactly Heisenberg's statement of the Uncertainty Principle. It lies at the heart of quantum theory, but we should be quite clear that it is not in itself a vague statement. It is a statement about our inability to track particles around with precision, and there is no more scope for quantum magic here than there is for Newtonian magic. What we have done in the last few pages is to derive Heisenberg's Uncertainty Principle from the fundamental rules of quantum physics as embodied in the rules for winding, shrinking and adding clocks. Indeed, its origin lies in our proposition that a particle can be anywhere in the Universe an instant after we measure its position. Our initial wild proposal that the particle can be anywhere and everywhere in the Universe has been tamed by the orgy of quantum interference, and the Uncertainty Principle is in a sense all that remains of the original anarchy.

There is something very important that we should say about how to interpret the Uncertainty Principle before we move on. We must not make the mistake of thinking that the particle is actually at some single specific place and that the spread in initial clocks really reflects some limitation in our understanding. If we thought that then we would not have been able to compute the Uncertainty Principle correctly, because we would not admit that we must take clocks from every possible point inside the initial cluster, transport them in turn to a distant point X and then add them all up. It was the act of doing this that gave us our result, i.e. we had to suppose that the particle arrives at X via a superposition of many possible routes. We will make use of Heisenberg's principle in some real-world examples later on. For now, it is satisfying that we have derived one of the key results of quantum theory using nothing more than some simple manipulations with imaginary clocks.

Let's stick a few numbers into the equations to get a better feel for things. How long will we have to wait for there to be a reasonable probability that a sand grain will hop outside a matchbox? Let's assume that the matchbox has sides of length 3 cm and that the sand grain weighs 1 microgram. Recall that the condition for there to be a reasonable probability of the sand grain hopping a given distance is given by

$$\frac{mx\Delta x}{ht} < 1$$

where Δx is the size of the matchbox. Let's calculate what t should be if we want the sand grain to jump a distance $x = 4$ cm, which would comfortably exceed the size of the matchbox. Doing a very simple bit of algebra, we find that

$$t > \frac{mx\Delta x}{h}$$

and sticking the numbers in tells us that t must be greater than approximately 10^{21} seconds. That is around 6×10^{13} years, which is over a thousand times the current age of the Universe. So it probably won't happen. Quantum mechanics is weird, but not weird enough to allow a grain of sand to hop unaided out of a matchbox.

To conclude this chapter, and launch ourselves into the next one, we will make one final observation. Our derivation of the Uncertainty Principle was based upon the configuration of clocks illustrated in Figure 4.4. In particular, we set up the initial cluster of clocks so that they all had hands of the same size and were all reading the same time. This specific arrangement corresponds to a particle initially at rest within a certain region of space – a sand grain in a matchbox, for example. Although we discovered that the particle will most likely not remain at rest, we also discovered that for large objects – and a grain of sand is very large indeed in quantum terms – this motion is completely undetectable. So there is some motion in our theory, but it is motion that is imperceptible for big enough objects. Obviously we are missing something rather important, because big things do actually move around, and remember that quantum theory is a theory of all things big and small. We must now address this problem: how can we explain motion?

5. Movement as an Illusion

In the previous chapter we derived Heisenberg's Uncertainty Principle by considering a particular initial arrangement of clocks – a small cluster of them, each with hands of the same size and pointing in the same direction. We discovered that this represents a particle that is approximately stationary, although the quantum rules imply that it jiggles around a little. We shall now set up a different initial configuration; we want to describe a particle in motion. In Figure 5.1, we've drawn a new configuration of clocks. Again it is a cluster of clocks, corresponding to a particle that is initially located in the vicinity of the clocks. The clock at position 1 reads 12 o'clock, as before, but the other clocks in the cluster are now all wound forwards by different amounts. We've drawn five clocks this time simply because it will help make the reasoning more transparent, although as before we are to imagine clocks in between the ones we have drawn – one for each point in space in the cluster. Let's apply the quantum rule as before and move these clocks to point X, a long way outside the cluster, to once again describe the many ways that the particle can hop from the cluster to X.

In a procedure that we hope is becoming more routine, let's take the clock from point 1 and propagate it to point X, winding it around as we go. It will wind around by an amount

$$W_1 = \frac{mx^2}{2ht}$$

Now let's take the clock from point 2 and propagate it to point X. It's a little bit further away, let's say a distance d further, so it will wind a bit more

$$W_2 = \frac{m(x+d)^2}{2ht}$$

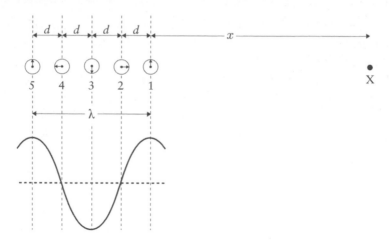

Figure 5.1: The initial cluster (illustrated by the clocks marked 1 to 5) is made up of clocks that all read different times – they are all shifted by three hours relative to their neighbours. The lower part of the figure illustrates how the time on the clocks varies through the cluster.

This is exactly what we did in the previous chapter, but perhaps you can already see that something different will happen for this new initial configuration of clocks. We set things up such that clock 2 was initially wound *forwards* by three hours relative to clock 1 – from 12 o'clock to 3 o'clock. But in carrying clock 2 to point X, we have to wind it *backwards* by a little more than clock 1, corresponding to the extra distance d that it has to travel. If we arrange things so that the initial forward wind of clock 2 is exactly the same as the extra backward wind it gets when travelling to X, then it will arrive at X *showing exactly the same time* as clock 1. This will mean that, far from cancelling out, it will add to clock 1 to make a larger clock, which in turn means that there will be a high probability that the particle will be found at X. This is a completely different situation from the orgy of quantum interference that occurred when we began with all the clocks reading the same time. Let's now consider clock 3, which we have wound forwards six hours relative to clock 1. This clock has to travel an extra distance $2d$ to make it to point X and again, because of the offset in time, this clock will arrive pointing to

12 o'clock. If we set all the offsets in the same manner, then this will happen right across the cluster and all of the clocks will add together constructively at X.

This means that there will be a high probability that the particle will be found at the point X at some later time. Clearly point X is special because it is that particular point where all the clocks from the cluster conspire to read the same time. But point X is not the only special point – all points to the left of X for a distance equal to the length of the original cluster also share the same property that the clocks add together constructively. To see this, notice that we could take clock 2 and transport it to a point a distance d to the left of X. This would correspond to moving it a distance x, which is exactly the same distance that we moved clock 1 when we moved it to X. We could then transport clock 3 to this new point through a distance $x + d$, which is exactly the same distance that we previously moved clock 2. These two clocks should therefore read the same time when they arrive and add together. We can keep on doing this for all the clocks in the cluster, but only until we reach a distance to the left of X equal to the original cluster size. Outside of this special region, the clocks largely cancel out because they are no longer protected from the usual orgy of quantum interference.[1] The interpretation is clear: the cluster of clocks moves, as illustrated in Figure 5.2.

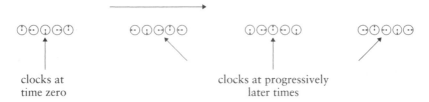

clocks at
time zero

clocks at progressively
later times

Figure 5.2. The cluster of clocks moves at constant speed to the right. This is because the original cluster had its clocks wound relative to each other as described in the text.

1. You might like to check this explicitly for yourself.

This is a fascinating result. By setting up the initial cluster using offset clocks rather than clocks all pointing in the same direction, we have arrived at the description of a moving particle. Intriguingly, we can also make a very important connection between the offset clocks and the behaviour of waves.

Remember that we were motivated to introduce the clocks back in Chapter 2 in order to explain the wave-like behaviour of particles in the double-slit experiment. Look back at Figure 3.3 on page 35, where we sketched an arrangement of clocks that describes a wave. It is just like the arrangement of the clocks in our moving cluster. We've sketched the corresponding wave below the cluster in Figure 5.1 using exactly the same methodology as before: 12 o'clock represents the peak of the wave, 6 o'clock represents the trough and 3 o'clock and 9 o'clock represent the places where the wave height is zero.

As we might have anticipated, it appears that the representation of a moving particle has something to do with a wave. The wave has a wavelength, and this corresponds to the distance between clocks showing identical times in the cluster. We've also drawn this on the figure, and labelled it λ.

We can now work out how far the point X should be away from the cluster in order for adjacent clocks to add constructively. This will lead us to another very important result in quantum mechanics, and make the connection between quantum particles and waves much clearer. Time for a bit more mathematics.

First, we need to write down the extra amount by which clock 2 is wound relative to clock 1 because it has further to travel to point X. Using the results on page 75, this is

$$W_2 - W_1 = \frac{m(x + d)^2 - mx^2}{2ht} \simeq \frac{mxd}{ht}$$

Again, you may be able to work this out for yourself by multiplying out the brackets and throwing away the d^2 bits because d, the distance between the clocks, is very small compared to x, the distance to point X a long way away from the original cluster.

It is also straightforward to write down the criterion for the clocks to read the same time; we want the extra amount of winding due to the propagation of clock 2 to be exactly cancelled by the extra forward wind we gave it initially. For the example shown in Figure 5.1, the extra wind for clock 2 is $\frac{1}{4}$, because we've wound the clock forward by a quarter of a turn. Similarly, clock 3 has a wind of $\frac{1}{2}$, because we've wound it around $\frac{1}{2}$ a turn. In symbols, we can express the fraction of one full wind between two clocks quite generally as d/λ, where d is the distance between the clocks and λ is the wavelength. If you can't quite see this, just think of the case for which the distance between two clocks is equal to the wavelength. Then $d = \lambda$, and therefore $d/\lambda = 1$, which is one full wind, and both clocks will read the same time.

Bringing this all together, we can say that for two adjacent clocks to read the same time at point X we require the extra amount of wind we put into the initial clock to be equal to the extra amount of wind due to the difference in propagation distance:

$$\frac{mxd}{ht} = \frac{d}{\lambda}$$

We can simplify this, as we've done before, by noticing that mx/t is the momentum of the particle, p. So with a little bit of rearrangement, we get

$$p = \frac{h}{\lambda}$$

This result is important enough to warrant a name, and it is called the de Broglie equation because it was first proposed in September 1923 by the French physicist Louis de Broglie. It is important because it associates a wavelength with a particle of a known momentum. In other words it expresses an intimate link between a property usually associated with particles – momentum – and a property usually associated with waves – wavelength. In this way, the wave-particle duality of quantum mechanics has emerged from our manipulations with clocks.

The de Broglie equation constituted a huge conceptual leap. In his original paper, he wrote that a 'fictitious associated wave' should be assigned to all particles, including electrons, and that a stream of electrons passing through a slit 'should show diffraction phenomena'.[2] In 1923, this was theoretical speculation, because Davisson and Germer did not observe an interference pattern using beams of electrons until 1927. Einstein made a similar proposal to de Broglie's, using different reasoning, at around the same time, and these two theoretical results were the catalyst for Schrödinger to develop his wave mechanics. In the last paper before he introduced his eponymous equation, Schrödinger wrote: 'That means nothing else but taking seriously the de Broglie–Einstein wave theory of moving particles.'

We can gain a little more insight into the de Broglie equation by looking at what happens if we decrease the wavelength, which would correspond to increasing the amount of winding between adjacent clocks. In other words, we will reduce the distance between clocks reading the same time. This means that we would then have to increase the distance x to compensate for the decrease in λ. In other words, point X needs to be further away in order for the extra winding to be 'undone'. That corresponds to a faster-moving particle: smaller wavelength corresponds to larger momentum, which is exactly what the de Broglie equation says. It is a lovely result that we have managed to 'derive' ordinary motion (because the cluster of clocks moves smoothly in time) starting from a static array of clocks.

Wave Packets

We would now like to return to an important issue that we skipped over earlier in the chapter. We said that the initial cluster moves in its entirety to the vicinity of point X, but only roughly maintains its

2. 'Diffraction' is a word used to describe a particular type of interference, and it is characteristic of waves.

original configuration. What did we mean by that rather imprecise statement? The answer provides a link back to the Heisenberg Uncertainty Principle, and delivers further insight.

We have been describing what happens to a cluster of clocks, which represents a particle that can be found somewhere within a small region of space. That's the region spanned by our five clocks in Figure 5.1. A cluster like this is referred to as a wave packet. But we have already seen that confining a particle to some region in space has consequences. We cannot prevent a localized particle from getting a Heisenberg kick (i.e. its momentum is uncertain because it is localized), and as time passes this will lead to the particle 'leaking out' of the region within which it was initially located. This effect was present for the case where the clocks all read the same time and it is present in the case of the moving cluster too. It will tend to spread the wave packet out as it travels, just as a stationary particle spreads out over time.

If we wait long enough, the wave packet corresponding to the moving cluster of clocks will have totally disintegrated and we'll lose any ability to predict where the particle actually is. This will obviously have implications for any attempt we might make to measure the speed of our particle. Let's see how this works out.

A good way to measure a particle's speed is to make two measurements of its position at two different times. We can then deduce the speed by dividing the distance the particle travelled by the time between the two measurements. Given what we've just said, however, this looks like a dangerous thing to do because if we make a measurement of the position of a particle too precisely then we are in danger of squeezing its wave packet, and that will change its subsequent motion. If we don't want to give the particle a significant Heisenberg kick (i.e. a significant momentum because we make Δx too small) then we must make sure that our position measurement is sufficiently vague. Vague is, of course, a vague term, so let's make it less so. If we use a particle detection device that is capable of detecting particles to an accuracy of 1 micrometre and our wave packet has a width of 1 nanometre, then the detector won't have much impact on the particle at all. An experimenter reading out the detector might be very happy with a resolution of 1 micron but, from the electron's perspective, all the detector did was report back to the experimenter that the particle is in some huge box, a thousand times bigger than the actual wave packet. In this case, the Heisenberg kick induced by the measurement process will be very small compared to that induced by the finite size of the wave packet itself. That's what we mean by 'sufficiently vague'.

We've sketched the situation in Figure 5.3 and have labelled the initial width of the wave packet d and the resolution of our detector Δ. We've also drawn the wave packet at a later time; it's a little broader and has a width d', which is bigger than d. The peak of the wave packet has travelled a distance L over some time interval t at a speed v. Apologies if that particular flourish of formality reminds you of your long-forgotten school days sitting behind a stained and eroded wooden bench listening to a science teacher's voice fading into the half-light of a late winter's afternoon as you slide into an inappropriate nap. We are covering ourselves in chalk dust for good reason, and it is our hope that the conclusion of this section will jolt you back to consciousness more effectively than the flying board dusters of your youth.

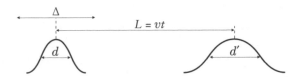

Figure 5.3. A wave packet at two different times. The packet moves to the right and spreads out as time advances. The packet moves because the clocks that constitute it are wound around relative to each other (de Broglie) and it spreads out because of the Uncertainty Principle. The shape of the packet is not very important but, for completeness, we should say that where the packet is large the clocks are large, and where it is small the clocks are small.

Back in the metaphorical science lab, with renewed vigour, we are trying to measure the speed v of the wave packet by making two measurements of its position at two different times. This will give us the distance L that the wave packet has travelled in a time t. But our detector has a resolution Δ, so we won't be able to pin down L exactly. In symbols, we can say that the measured speed is

$$v = \frac{L \pm \Delta}{t}$$

where the combined plus or minus sign is there simply to remind us that, if we actually make the two position measurements, we will generally not always get L but instead 'L plus a bit' or 'L minus a bit', where the 'bit' is due to the fact we agreed not to make a very accurate measurement of the particle's position. It is important to bear in mind that L is not something we can actually measure: we always measure a value somewhere in the range $L \pm \Delta$. Remember also that we need Δ to be much larger than the size of the wave packet otherwise we will squeeze the particle and that will disrupt it.

Let's rewrite the last equation very slightly so that we can better see what's going on:

$$v = \frac{L}{t} \pm \frac{\Delta}{t}$$

It seems that if we take t to be very large then we will get a measurement of the speed $v = L/t$ with a very tiny spread, because we can choose to wait around for a very long time, making t as large as we like and consequently Δ/t as small as we like whilst still keeping Δ comfortably large. This looks like we have a nice way to make an arbitrarily precise measurement of the particle's speed without disturbing it at all; just wait for a huge amount of time between the first and the second measurements. This makes perfect intuitive sense. Imagine you are measuring the speed of a car driving along a road. If you measure how far it has travelled in one minute, you will tend to get a much more precise measurement of its speed than if you measure how far it travelled in one second. Have we dodged Heisenberg?

Of course not – we have forgotten to take something into

account. The particle is described by a wave packet that spreads out as time passes. Given enough time, the spreading out will completely wash out the wave packet and that means the particle could be anywhere. This will increase the range of values we get in our measurement of L and spoil our ability to make an arbitrarily accurate measurement of its speed.

For a particle described by a wave packet, we are ultimately still bound by the Uncertainty Principle. Because the particle is initially confined in a region of size d, Heisenberg informs us that the particle's momentum is correspondingly blurred out by an amount equal to h/d.

There is therefore only one way we can build a configuration of clocks to represent a particle that travels with a definite momentum – we must make d, the size of the wave packet, very large. And the larger we make it, the smaller the uncertainty in its momentum will be. The lesson is clear: a particle of well-known momentum is described by a large cluster of clocks.[3] To be precise, a particle of absolutely definite momentum will be described by an infinitely long cluster of clocks, which means an infinitely long wave packet.

We have just argued that a finite-size wave packet does not correspond to a particle with a definite momentum. This means that if we measured the momentum of very many particles, all described by exactly the same initial wave packet, then we would not get the same answer each time. Instead we would get a spread of answers and it does not matter how brilliant we are at experimental physics, that spread cannot be made smaller than h/d.

We can therefore say that a wave packet describes a particle that is travelling with a range of momenta. But the de Broglie equation implies that we can just substitute the word 'wavelengths' for 'momenta' in the last sentence, because a particle's momentum is

---

3. Of course if d is very large then one might wonder how we can even measure the momentum. That concern is sidestepped by ensuring that no matter how big d is, L is much bigger than it.

associated with a wave of definite wavelength. This in turn means that a wave packet must be made up of many different wavelengths. Likewise, if a particle is described by a wave with a definite wavelength then that wave must necessarily be infinitely long. It sounds like we are being pushed to conclude that a small wave packet is made up of many infinitely long waves of different wavelengths. We are indeed being pushed down this route, and what we are describing is very familiar to mathematicians, physicists and engineers alike. This is an area of mathematics known as Fourier analysis, named after the French mathematical physicist Joseph Fourier.

Fourier was a colourful man. Amongst his many notable achievements, he was Napoleon's governor of Lower Egypt and the discoverer of the greenhouse effect. He apparently enjoyed wrapping himself up in blankets, which led to his untimely demise one day in 1830 when, tightly wrapped, he fell down his own stairs. His key paper on Fourier analysis addressed the subject of heat transfer in solids and was published in 1807, although the basic idea can be traced back much earlier.

Fourier showed that any wave at all, of arbitrarily complex shape and extent, can be synthesized by adding together a number of sine waves of different wavelengths. The point is best illustrated through pictures. In Figure 5.4 the dotted curve is made by adding together the first two sine waves in the lower graphs. You can almost do the addition in your head – the two waves are both at maximum height in the centre, and so they add together there, whilst they tend to cancel each other out at the ends. The dashed curve is what happens if we add together all four of the waves illustrated in the lower graphs – now the peak in the centre is becoming more pronounced. Finally, the solid curve shows what happens when we add together the first ten waves, i.e. the four shown plus six more of progressively decreasing wavelength. The more waves we add in to the mix, the more detail we can achieve in the final wave. The wave packet in the upper graph could describe a localized particle, rather like the wave packet illustrated in Figure 5.3. In this way it really is possible

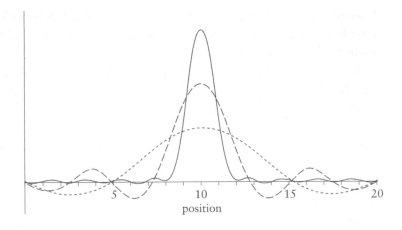

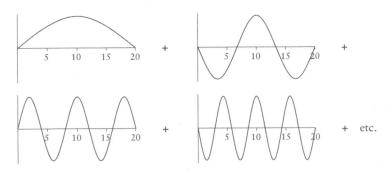

Figure 5.4. Upper graph: Adding together several sine waves to synthesize a sharply peaked wave packet. The dotted curve contains fewer waves than the dashed one, which in turn contains fewer than the solid one. Lower graphs: The first four waves used to build up the wave packets in the upper graph.

to synthesize a wave of any shape at all – it is all achieved by adding together simple sine waves.

The de Broglie equation informs us that each of the waves in the lower graphs of Figure 5.4 corresponds to a particle with a definite momentum, and the momentum increases as the wavelength decreases. We are beginning to see why it is that if a particle is described by a localized cluster of clocks then it must necessarily be made up of a range of momenta.

To be more explicit, let's suppose that a particle is described by the cluster of clocks represented by the solid curve in the upper graph in Figure 5.4.[4] We have just learnt that this particle can also be described by a series of much longer clusters of clocks: the first wave in the lower graphs plus the second wave in the lower graphs, plus the third wave in the lower graphs, and so on. In this way of thinking, there are several clocks at each point (one from each long cluster), which we should add together to produce the single clock cluster represented in upper graph of Figure 5.4. The choice of how to think about the particle is really 'up to you'. You can think of it as being described by one clock at each point, in which case the size of the clock immediately lets you know where the particle is likely to be found, i.e. in the vicinity of the peak in the upper graph of Figure 5.4. Alternatively, you can think of it as being described by a number of clocks at each point, one for each possible value of the momentum of the particle. In this way we are reminding ourselves that the particle localized in a small region does not have a definite momentum. The impossibility of building a compact wave packet from a single wavelength is an evident feature of Fourier's mathematics.

This way of thinking provides us with a new perspective on Heisenberg's Uncertainty Principle. It says that we cannot describe a particle in terms of a localized cluster of clocks using clocks corresponding to waves of a single wavelength. Instead, to get the clocks to cancel outside the region of the cluster, we must necessarily mix in different wavelengths and hence different momenta. So, the price we pay for localizing the particle to some region in space is to admit we do not know what its momentum is. Moreover, the more we restrict the particle, the more waves we need to add in and

---

4. Recall that when we draw pictures of waves, they are really a convenient way of picturing what the projections of the clock hands in the 12 o'clock direction are.

the less well we know its momentum. This is exactly the content of the Uncertainty Principle, and it is very satisfying to have found a different way of reaching the same conclusion.[5]

To close this chapter we want to spend a little more time with Fourier. There is a very powerful way of picturing quantum theory that is intimately linked to the ideas we have just been discussing. The important point is that any quantum particle, whatever it is doing, is described by a wavefunction. As we've presented it so far, the wavefunction is simply the array of little clocks, one for each point in space, and the size of the clock determines the probability that the particle will be found at that point. This way of representing a particle is called the 'position space wavefunction' because it deals directly with the possible positions that a particle can have. There are, however, many ways of representing the wavefunction mathematically, and the little clocks in space version is only one of them. We touched on this when we said it is possible to think of the particle as also being represented by a sum over sine waves. If you ponder this point for a moment, you should realize that specifying the complete list of sine waves actually provides a complete description of the particle (because by adding together these waves we can obtain the clocks associated with the position space wavefunction). In other words, if we specify exactly which sine waves are needed to build a wave packet, and exactly how much of each sine wave we need to add in to get the shape just right, then we will have a different but entirely equivalent description of the wave packet. The neat thing is that any sine wave can itself be described by a single imaginary clock: the size of the clock encodes the maximum height of the wave and the phase of the wave at some point can be represented by the time that the clock reads. This means that we can choose to represent a particle not by clocks in space but by an alternative list of clocks, one for each possible value of the particle's momentum. This description is just as economical as the 'clocks in space' descrip-

5. This way of arriving at the Uncertainty Principle did, however, rely on the de Broglie equation in order to link the wavelength of a clock wave to its momentum.

tion, and instead of making explicit where the particle is likely to be found we are instead making explicit what values of momentum the particle is likely to have. This alternative array of clocks is known as the momentum space wavefunction and it contains exactly the same information as the position space wavefunction.[6]

This might sound very abstract, but you may well use technology based on Fourier's ideas every day, because the decomposition of a wave into its component sine waves is the foundation of audio and video compression technology. Think about the sound waves that make up your favourite tune. This complicated wave can, as we have just learnt, be broken down into a series of numbers that give the relative contributions of each of a large number of pure sine waves to the sound. It turns out that, although you may need a vast number of individual sine waves to reproduce the original sound wave exactly, you can in fact throw a lot of them away without compromising the perceived audio quality at all. In particular, the sine waves that contribute to sound waves that humans can't hear are not kept. This vastly reduces the amount of data needed to store an audio file – hence your mp3 player doesn't need to be too large.

We might also ask what possible use could this different and even more abstract version of the wavefunction be? Well, think of a particle represented, in position space, by a single clock. This describes a particle located at a certain place in the Universe; the single point where the clock sits. Now think of a particle represented by a single clock, but this time in momentum space. This represents a particle with a single, definite momentum. Describing such a particle using the position space wavefunction would, in contrast, require an infinite number of equally sized clocks, because according to the Uncertainty Principle, a particle with a definite momentum can be found anywhere. As a result, it is sometimes simpler to perform calculations directly in terms of the momentum space wavefunction.

6. In the jargon, the momentum space wavefunctions that correspond to particles with definite momentum are known as momentum eigenstates, after the German word *eigen*, meaning 'characteristic'.

In this chapter, we have learnt that the description of a particle in terms of clocks is capable of capturing what we ordinarily call 'movement'. We have learnt that our perception that objects move smoothly from point to point is, from the perspective of quantum theory, an illusion. It is closer to the truth to suppose that particles move from A to B via all possible paths. Only when we add together all of the possibilities does motion as we perceive it emerge. We have also seen explicitly how the clock description manages to encode the physics of waves, even though we only ever deal with point-like particles. It is time now to really exploit the similarity with the physics of waves as we tackle the important question: how does quantum theory explain the structure of atoms?

6.   The Music of the Atoms

The interior of an atom is a strange place. If you could stand on a proton and gaze outwards into inter-atomic space, you would see only void. The electrons would still be imperceptibly tiny even if they approached close enough for you to touch them, which they very rarely would. The proton is around 10^{-15} m in diameter, 0.000000000000001 metres, and is a quantum colossus compared to the electrons. If you stand on your proton at the edge of England on the White Cliffs of Dover, the fuzzy edge of the atom lies somewhere amongst the farms of northern France. Atoms are vast and empty, which means the full-size you is vast and empty too. Hydrogen is the simplest atom, comprising a single proton and a single electron. The electron, vanishingly small as far as we can tell, might seem to have a limitless arena within which to roam, but this is not true. It is bound to its proton, trapped by their mutual electromagnetic attraction, and it is the size and shape of this generous prison that gives rise to the characteristic barcode rainbow of light meticulously documented in the *Handbuch der Spectroscopie* by our old friend and dinner-party guest Professor Kayser.

We are now in a position to apply the knowledge we have accumulated so far to the question that so puzzled Rutherford, Bohr and others in the early decades of the twentieth century: what exactly is going on inside an atom? The problem, if you recall, was that Rutherford discovered that the atom is in some ways like a miniature solar system, with a dense nucleus Sun at the centre and electrons as planets sweeping around in distant orbits. Rutherford knew that this model couldn't be right, because electrons in orbit around a nucleus should continually emit light. The result should be catastrophic for the atom, because if the electron continually emits light then it must lose energy and spiral inwards on an inevitable collision course with the proton. This, of course, doesn't happen. Atoms tend to be stable things, so what is wrong with this picture?

This chapter marks an important stage in the book, because it is the first time that our theory is to be used to explain real-world phenomena. All our hard work to this point has been concerned with getting the essential formalism worked out so that we have a way to think about a quantum particle. Heisenberg's Uncertainty Principle and the de Broglie equation represent the pinnacle of our achievements, but in the main we have been modest, thinking about a universe containing just one particle. It is now time to show how quantum theory impacts on the everyday world in which we live. The structure of atoms is a very real and tangible thing. You are made of atoms: their structure is your structure, and their stability is your stability. It would not be unduly hyperbolic to say that understanding the structure of atoms is one of the necessary conditions for understanding our Universe as a whole.

Inside a hydrogen atom, the electron is trapped in a region surrounding the proton. We are going to start by imagining that the electron is trapped in some sort of box, which is not very far from the truth. Specifically, we'll investigate to what extent the physics of an electron trapped inside a tiny box captures the salient features of a real atom. We are going to proceed by exploiting what we learnt in the previous chapter about the wave-like properties of quantum particles, because, when it comes to describing atoms, the wave picture really simplifies things and we can make a good deal of progress without having to worry about shrinking, winding and adding clocks. Always bear in mind, though, that the waves are a convenient shorthand for what is going on 'under the bonnet'.

Because the framework we've developed for quantum particles is extremely similar to that used in the description of water waves, sound waves or the waves on a guitar string, we'll think first about how these more familiar material waves behave when they are confined in some way.

Generally speaking, waves are complicated things. Imagine jumping into a swimming pool full of water. The water will slosh around

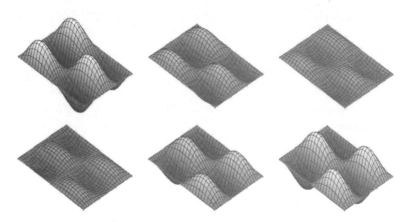

Figure 6.1. Six successive snapshots of a standing wave in a tank of water. The time advances from the top left to the bottom right.

all over the place, and it would seem to be futile to try to describe what is going on in any simple fashion. Underlying the complexity, however, there is hidden simplicity. The key point is that the water in a swimming pool is confined, which means that all the waves are trapped inside the pool. This gives rise to a phenomenon known as 'standing waves'. The standing waves are hidden away in the mess when we disturb the pool by jumping into it, but there is a way to make the water move so that it oscillates in the regular, repeating patterns of the standing waves. Figure 6.1 shows how the water surface looks when it is undergoing one such oscillation. The peaks and troughs rise and fall, but most importantly they rise and fall in exactly the same place. There are other standing waves too, including one where the water in the middle of the tank rises and falls rhythmically. We do not usually see these special waves because they are hard to produce, but the key point is that any disturbance of the water at all – even the one we caused by our inelegant dive and subsequent thrashing around – can be expressed as some combination or other of the different standing waves. We've seen this type of behaviour before; it is a direct generalization of Fourier's ideas that we encountered in the last chapter. There, we saw that any wave packet can be built up out of a combination of waves each of definite wavelength. These special waves, representing particle states of definite momentum, are sine waves. In the case of confined water waves, the idea generalizes so that any disturbance can always be described using some combination of standing waves. We'll see later in this chapter that standing waves have an important interpretation in quantum theory, and in fact they hold the key to understanding the structure of atoms. With this in mind, let's explore them in a little more detail.

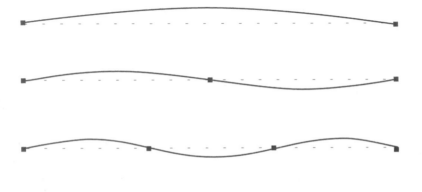

Figure 6.2. The three longest wavelength waves that can fit on a guitar string. The longest wavelength (at the top) corresponds to the lowest harmonic (fundamental) and the others correspond to the higher harmonics (overtones).

Figure 6.2 shows another example of standing waves in Nature: three of the possible standing waves on a guitar string. On plucking a guitar string, the note we hear is usually dominated by the standing wave with the largest wavelength – the first of the three waves shown in the figure. This is known in both physics and music as the 'lowest harmonic' or 'fundamental'. Other wavelengths are usually present too, and they are known as overtones or higher harmonics. The other waves in the figure are the two longest-wavelength overtones. The guitar is a nice example because it's simple enough to see why a guitar string can only vibrate at these special wavelengths. It is because it is held fixed at both ends – by the guitar bridge at one end and your finger pressing against a fret at the other. This means that the string cannot move at these two points, and this determines the allowed wavelengths. If you play the guitar, you'll know this physics instinctively; as you move your fingers up the fret board towards the bridge, you decrease the length of the string and therefore force it to vibrate with shorter and shorter wavelengths, corresponding to higher-pitched notes.

The lowest harmonic is the wave that has only two stationary points, or 'nodes'; it moves everywhere except at the two fixed ends. As you can see from the figure, this note has a wavelength of twice the length of the string. The next smallest wavelength is equal to the length of the string, because we can fit another node in the centre. Next, we can get a wave with wavelength equal to $\frac{2}{3}$ times the length of the string, and so on.

In general, just as in the case of the water confined in a swimming pool, the string will vibrate in some combination of the different possible standing waves, depending on how it is plucked. The actual shape of the string can always be obtained by adding together the standing waves corresponding to each of the harmonics present. The harmonics and their relative sizes give the sound its characteristic tone. Different guitars will have different distributions of harmonics and therefore sound different, but a middle C (a pure harmonic) on one guitar is always the same as a middle C on another. For the guitar, the shape of the standing waves is very simple: they are pure sine waves whose wavelengths are fixed by the length of the string. For the swimming pool, the standing waves are more complicated, as shown in Figure 6.1, but the idea is exactly the same.

You may be wondering why these special waves are called 'standing waves'. It is because the waves do not change their shape. If we take two snapshots of a guitar string vibrating in a standing wave, then the two pictures will only differ in the overall size of the wave. The peaks will always be in the same place, and the nodes will always be in the same place because they are fixed by the ends of the string or, in the case of the swimming pool, by the sides of the pool. Mathematically, we could say that the waves in the two snapshots differ only by an overall multiplicative factor. This factor varies periodically with time, and expresses the rhythmical vibration of the string. The same is true for the swimming pool in Figure 6.1, where each snapshot is related to the others by an overall multiplicative factor. For example, the last snapshot can be obtained from the first by multiplying the wave height at every point by minus one.

In summary, waves that are confined in some way can always be expressed in terms of standing waves (waves that do not change their shape) and, as we have said, there are very good reasons for devoting so much time to understanding them. At the top of the list is the fact that standing waves are *quantized*. This is very clear for the standing waves on a guitar string: the fundamental has a wavelength of twice the length of the string, and the next longest allowed wavelength is equal to the length of the string. There is no standing wave with a wavelength in between these two and so we can say that the allowed wavelengths on a guitar string are quantized.

Standing waves therefore make manifest the fact that something gets quantized when we trap waves. In the case of a guitar string, it is clearly the wavelength. For the case of an electron inside a box, the quantum waves corresponding to the electron will also be trapped, and by analogy we should expect that only certain standing waves will be present in the box, and therefore that something will be quantized. Other waves simply cannot exist, just as a guitar string doesn't play all the notes in an octave at the same time no matter how it is plucked. And just as for the sound of a guitar, the general state of the electron will be described by a blend of standing waves. These quantum standing waves are starting to look very interesting, and, encouraged by this, let's start our analysis proper.

To make progress, we must be specific about the shape of the box inside which we place our electron. To keep things simple, we'll suppose that the electron is free to hop around inside a region of size L, but that it is totally forbidden from wandering outside this region. We do not need to say how we intend to forbid the electron from wandering – but if this is supposed to be a simplified model of an atom then we should imagine that the force exerted by the positively charged nucleus is responsible for its confinement. In the jargon, this is known as a 'square well potential'. We've sketched the situation in Figure 6.3, and the reason for the name should be obvious.

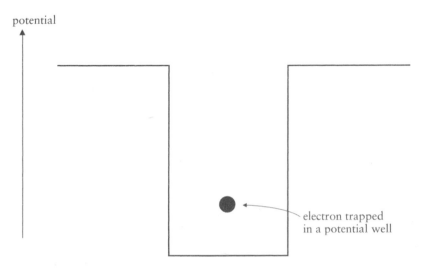

Figure 6.3. An electron trapped in a square well potential.

The idea of confining a particle in a potential is a very important one that we'll use again, so it will be useful to make sure we understand exactly what it means. How do we actually trap particles? That is quite a sophisticated question; to get to the bottom of it we'll need to learn about how particles interact with other particles, which we will do in Chapter 10. Nevertheless, we can make progress provided we don't ask too many questions.

The ability 'not to ask too many questions' is a necessary skill in physics because we have to draw the line somewhere in order to answer any questions at all; no system of objects is perfectly isolated. It seems reasonable that if we want to understand how a microwave oven works, we don't need to worry about any traffic passing by outside. The traffic will have a tiny influence on the operation of the oven. It will induce vibrations in the air and ground which will shake the oven a little bit. There may also be stray magnetic fields that influence the internal electronics of the oven, no matter how well they are shielded. It is possible to make mistakes in ignoring things because there might be some crucial detail that we miss. If this is the case, we'll simply get the wrong answer and have to reconsider our assumptions. This is very important, and goes to the heart of the success of science; all assumptions are ultimately validated or negated by experiment. Nature is the arbiter, not human intuition. Our strategy here is to ignore the details of the mechanism that traps the electron and model it by something called a potential. The word 'potential' really just means 'an effect on the particle due to some physics or other that I will not bother to explain in detail'. We will bother to describe in detail how particles interact later on, but for now we'll talk in the language of potentials. If this sounds a bit cavalier, let us give an example to illustrate how potentials are used in physics.

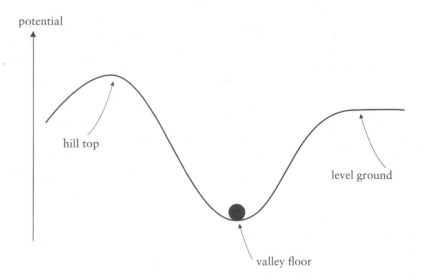

Figure 6.4. A ball sitting on a valley floor. The height of the ground above sea-level is directly proportional to the potential that the particle experiences when it rolls around.

Figure 6.4 illustrates a ball trapped in a valley. If we give the ball a kick then it can roll up the valley, but only so far, and then it will roll back down again. This is an excellent example of a particle trapped by a potential. In this case, the Earth's gravitational field generates the potential and a steep hill makes a steep potential. It should be clear that we could calculate the details of how a ball rolls around in a valley without knowing the precise details of how the valley floor interacts with the ball – for this we'd have to know about the theory of quantum electrodynamics. If it turned out that the details of the inter-atomic interactions between the atoms in the ball and the atoms in the valley floor affected the motion of the ball too much, then the predictions we make would be wrong. In fact, the inter-atomic interactions are important because they give rise to friction, but we can also model this without getting into Feynman diagrams. But we digress.

This example is very tangible because we can literally see the shape of the potential[1]. However, the idea is more general and works for potentials other than those created by gravity and valleys. An example is the electron trapped in a square well. Unlike the case of the ball in a valley, the height of the walls is not the actual height of anything; rather it represents how fast the electron needs to be moving before it can escape from the well. For the case of a valley, this would be analogous to rolling the ball so fast that it climbed up the walls and out of the valley. If the electron is moving slowly enough then the actual height of the potential won't matter much, and we can safely assume that the electron is confined to the interior of the well.

Let us now focus on the electron trapped inside a box described by a square well potential. Since it cannot escape from the box, the quantum waves must fall to zero at the edges of the box. The three possible quantum waves with the largest wavelengths are then

---

1. The fact that the gravitational potential exactly maps the terrain is because, in the vicinity of the Earth's surface, the gravitational potential is proportional to the height above the ground.

entirely analogous to the guitar-string waves illustrated in Figure 6.2: the longest possible wavelength is twice the size of the box, $2L$; the next longest wavelength is equal to the size of the box, L; and the next again has a wavelength of $2L/3$. Generally, we can fit electron waves with wavelength $2L/n$ in the box where $n = 1, 2, 3, 4$, etc.

Specifically for the square box, therefore, the electron waves are precisely the same shape as the waves on a guitar string; they are sine waves with a very particular set of allowed wavelengths. Now we can go ahead and invoke the de Broglie equation from the last chapter to relate the wavelength of these sine waves to the momentum of the electron via $p = h/\lambda$. In which case, the standing waves describe an electron that is only allowed to have certain momenta, given by the formula $p = nh/(2L)$, where all we did here was to insert the allowed wavelengths into the de Broglie equation.

And so it is that we have demonstrated that the momentum of our electron is quantized in a square well. This is a big deal. However, we do need to take care. The potential in Figure 6.3 is a special case, and for other potentials the standing waves are not generally sine waves. Figure 6.5 shows a photograph of the standing waves on a drum. The drum skin is sprinkled with sand, which collects at the nodes of the standing wave. Because the boundary enclosing the vibrating drum skin is circular, rather than square, the standing waves are no longer sine waves.[2] This means that, as soon as we move to the more realistic case of an electron trapped by a proton, its standing waves will likewise not be sine waves. In turn this means that the link between wavelength and momentum is lost. How, then, are we to interpret these standing waves? What is it that is generally quantized for trapped particles, if it isn't their momentum?

We can get the answer by noticing that in the square well potential, if the electron's momentum is quantized, then so too is its energy. That is a simple observation and appears to contain no important new information, since energy and momentum are simply related to each other. Specifically, the energy $E = p^2/2m$,

---

2. They are in fact described by Bessel functions.

Figure 6.5. A vibrating drum covered in sand. The sand collects at the nodes of the standing waves.

where p is the momentum of the trapped electron and m is its mass.[3] This is not such a pointless observation as it might appear, because, for potentials that are not as simple as the square well, each standing wave *always* corresponds to a particle of definite energy.

The important difference between energy and momentum emerges because $E = p^2/2m$ is only true when the potential is flat in the region where the particle can exist, allowing the particle to move freely, like a marble on a table top or, more to the point, an electron in a square well. More generally, the particle's energy will not be equal to $E = p^2/2m$; rather it will be the sum of the energy due to its motion and its potential energy. This breaks the simple link between the particle's energy and its momentum.

3. This is obtained using the fact that the energy is equal to $\frac{1}{2}mv^2$ and $p = mv$. These equations do get modified by Special Relativity but the effect is small for an electron inside a hydrogen atom.

We can illustrate this point by thinking again about the ball in a valley, shown in Figure 6.4. If we start with the ball resting happily on the valley floor, then nothing happens.[4] To make it roll up the side of the valley, we'd have to give it a kick, which is equivalent to saying that we need to add some energy to it. The instant after we kick the ball, all of its energy will be in the form of kinetic energy. As it climbs the side of the valley, the ball will slow down until, at some height above the valley floor, it will come to a halt before rolling back down again and up the other side. At the moment it stops, high up the valley side, it has no kinetic energy, but the energy hasn't just magically vanished. Instead, all of the kinetic energy has been changed into potential energy, equal to mgh, where g is the acceleration due to gravity at the Earth's surface and h is the height of the ball above the valley floor. As the ball starts to roll back down into the valley, this stored potential energy is gradually converted back into kinetic energy as the ball speeds up again. So as the ball rolls from one side of the valley to the other, the total energy remains constant, but it periodically switches between kinetic and potential. Clearly, the ball's momentum is constantly changing, but its energy remains constant (we have pretended that there is no friction to slow the ball down. If we did include it then the total energy would still be constant but only after including the energy dissipated via friction).

We are now going to explore the link between standing waves and particles of definite energy in a different way, without appealing to the special case of the square well. We'll do this using those little quantum clocks.

First, notice that, if an electron is described by a standing wave at some instant in time, then it will be described by the same standing wave at some later time. By 'the same', we mean that the shape of the wave is unchanged, as was the case for the standing water wave in Figure 6.1. We don't, of course, mean that the wave does not change

---

4. This is a big ball and we don't need to worry about any quantum jiggling. But, if the thought crossed your mind, it is a good sign: your intuition is becoming quantized.

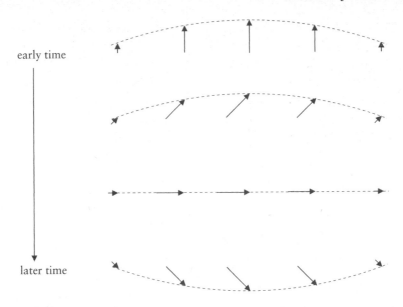

early time

later time

Figure 6.6: Four snapshots of a standing wave at successively later times. The arrows represent the clock hands and the dotted line is the projection onto the '12 o'clock' direction. The clocks all turn around in unison.

at all; the water height does change, but crucially the positions of the peaks and nodes do not. This allows us to figure out what the quantum clock description of a standing wave must look like, and it is illustrated in Figure 6.6 for the case of the fundamental standing wave. The clock sizes along the wave reflect the position of the peaks and nodes, and the clock hands sweep around together at the same rate. We hope you can see why we've drawn this particular pattern of clocks. The nodes must always be nodes, the peaks must always be peaks and they must always stay in the same place. This means that the clocks sitting in the vicinity of the nodes must *always* be very small, and the clocks representing the peaks must *always* have the longest hands. The only freedom we have, therefore, is to allow the clocks to sit where we put them and rotate in sync.

If we were following the methodology of the earlier chapters, we would now start from the configuration of clocks shown in the top row of Figure 6.6 and use the shrinking and turning rules to generate the bottom three rows at later times. This exercise in clock hopping is a hop too far for this book, but it can be done, and there is a nice twist because to do it correctly it is necessary to include the possibility that the particle 'bounces off the walls of the box' before hopping to its destination. Incidentally, because the clocks are bigger in the centre, we can immediately conclude that an electron described by this array of clocks is more likely to be found in the middle of the box than at the edges.

So, we have found that the trapped electron is described by an array of clocks that all whizz around at the same rate. Physicists don't usually talk like this, and musicians certainly don't; they both say that standing waves are waves of definite frequency.[5] High-frequency waves correspond to clocks that whizz around faster than the clocks of low-frequency waves. You can see this because, if a clock whizzes around faster, then the time it takes a peak to turn into a trough and then rise back again (represented by a single rotation of the clock hand) decreases. In terms of water waves, the high-frequency standing waves move up and down faster than the low-frequency ones. In music, a middle C is said to have a frequency of 262 Hz, which means that, on a guitar, the string vibrates up and down 262 times every second. The A above middle C has a frequency of 440 Hz, so it vibrates more rapidly (this is the agreed tuning standard for most orchestras and musical instruments across the world). As we've noted, however, it is only for pure sine waves that these waves of definite frequency also have definite wavelength. Generally speaking, *frequency* is the fundamental quantity that describes standing waves, and this sentence is probably a pun.

The million-dollar question, then, is 'What does it mean to speak of an electron of a certain frequency?' We remind you that these electron states are interesting to us because they are quantized and because an electron in one such state remains in that state for all

---

5. Actually, musicians probably don't say this either, and drummers definitely don't, because 'frequency' is a word with more than two syllables.

time (unless something enters the region of the potential and gives the electron a whack).

That last sentence is the big clue we need to establish the significance of 'frequency'. We encountered the law of energy conservation earlier in the chapter, and it is one of the few non-negotiable laws of physics. Energy conservation dictates that if an electron inside a hydrogen atom (or a square well) has a particular energy, then that energy *cannot* change until 'something happens'. In other words, an electron cannot spontaneously change its energy without a reason. This might sound uninteresting, but contrast this with the case of an electron that is known to be located at a point. As we know very well, the electron will leap off across the Universe in an instant, spawning an infinity of clocks. But the standing wave clock pattern is different. It keeps its shape, with all the clocks happily rotating away for ever unless something disturbs them. The unchanging nature of standing waves therefore makes them a clear candidate to describe an electron of definite energy.

Once we make the step of associating the frequency of a standing wave with the energy of a particle then we can exploit our knowledge of guitar strings to infer that higher frequencies must correspond to higher energies. That is because high frequency implies short wavelength (since short strings vibrate faster) and, from what we know of the special case of the square well potential, we can anticipate that a shorter wavelength corresponds to a higher-energy particle via de Broglie. The important conclusion, therefore, and all that really needs to be remembered for what follows, is that *standing waves describe particles of definite energy and the higher the energy the faster the clocks whizz round.*

In summary, we have deduced that when an electron is confined by a potential, its energy is quantized. In the physics jargon, we say that a trapped electron can only exist in certain 'energy levels'. The lowest energy the electron can have corresponds to its being described by the 'fundamental' standing wave alone,[6] and this energy

---

6. i.e. $n = 1$ in the case of the square well potential.

level is usually referred to as the 'ground state'. The energy levels corresponding to standing waves with higher frequencies are referred to as 'excited states'.

Let us imagine an electron of a particular energy, trapped in a square well potential. We say that it is 'sitting in a particular energy level' and its quantum wave will be associated with a single value of n (see page 100). The language 'sitting in a particular energy level' reflects the fact that the electron doesn't, in the absence of any external influence, do anything. More generally, the electron could be described by many standing waves at once, just as the sound of a guitar will be made up of many harmonics at once. This means that the electron will not in general have a unique energy.

Crucially, a measurement of the electron's energy must always reveal a value equal to that associated with one of the contributing standing waves. In order to compute the probability of finding the electron with a particular energy, we should take the clocks associated with the specific contribution to the total wavefunction coming from the corresponding standing wave, square them all up and add them all together. The resulting number tells us the probability that the electron is in this particular energy state. The sum of all such probabilities (one for each contributing standing wave) must add up to one, which reflects the fact that we will always find that the particle has an energy that corresponds to a specific standing wave.

Let's be very clear: an electron can have several different energies at the same time, and this is just as weird a statement as saying that it has a variety of positions. Of course, by this stage in the book this ought not to be such a shock, but it is shocking to our everyday sensibilities. Notice that there is a crucial difference between a trapped quantum particle and the standing waves in a swimming pool or on a guitar string. In the case of the waves on a guitar string, the idea that they are quantized is not at all weird, because the actual wave describing the vibrating string is simultaneously composed of many different standing waves, and all those waves physically contribute to the total energy of the wave. Because they can be mixed together in any way, the actual energy of the vibrating string can take on any value at all. For an electron trapped inside an atom, however, the relative contribution of each standing wave describes the probability that the electron will be found with that particular energy. The crucial difference arises because water waves are waves of water molecules but electron waves are most certainly not waves of electrons.

These deliberations have shown us that the energy of an electron inside an atom is quantized. This means that the electron is simply unable to possess any energy intermediate between certain allowed values. This is just like saying that a car can travel at 10 miles per hour or 40 miles per hour, but at no other speeds in between. Immediately, this fantastically bizarre conclusion offers us an explanation for why atoms do not continuously radiate light as the electron spirals into the nucleus. It is because there is no way for the electron to constantly shed energy, bit by bit. Instead, the only way it can shed any energy is to lose a whole chunk in one go.

We can also relate what we have just learnt to the observed properties of atoms, and in particular we can explain the unique colours of light they emit. Figure 6.7 shows the visible light emitted from the simplest atom, hydrogen. The light is composed of five distinct colours, a bright-red line corresponding to light with a wavelength of 656 nanometres, a light-blue line of wavelength 486 nanometres, and three other violet lines which fade away into the ultraviolet end of the spectrum. This series of coloured lines is known as the

400nm   486nm   656nm
(purple)   (blue)   (red)

Figure 6.7. The Balmer series for hydrogen: this is what happens when light from hydrogen gas is passed through a spectroscope.

Balmer series, after the Swiss mathematical physicist Johann Balmer, who wrote down a formula able to describe them in 1885. Balmer had no idea why his formula worked, because quantum theory was yet to be discovered – he simply expressed the regularity behind the pattern in a simple mathematical formula. But we can do better, and it is all to do with the allowed quantum waves that fit inside the hydrogen atom.

We know that light can be thought of as a stream of photons, each of energy $E = hc/\lambda$, where λ is the wavelength of the light.[7] The observation that atoms only emit certain colours of light therefore means that they only emit photons of very specific energies. We have also learnt that an electron 'trapped in an atom' can only possess certain very specific energies. It is a small step now to explain the long-standing mystery of the coloured light emitted from atoms: the different colours correspond to the emission of photons when electrons 'drop down' from one allowed energy level to another. This idea implies that the observed photon energies should always correspond to differences between a pair of allowed electron energies. This way of describing the physics nicely illustrates the value of expressing the state of the electron in terms of its allowed energies. If we had instead chosen to talk about the allowed values of the electron's momentum then the quantum nature would not be so apparent and we would not so easily conclude that the atom can only emit and absorb radiation at specific wavelengths.

The particle-in-a-box model of an atom is not accurate enough to allow us to compute the electron energies in a real atom, which is necessary to check this idea. But accurate calculations can be done if we model more accurately the potential in the vicinity of the proton that traps the electron. It is enough to say that these calculations confirm, without any shadow of doubt, that this really is the origin of those enigmatic spectral lines.

---

7. Incidentally, if you know that $E = cp$ for massless particles, which is a consequence of Einstein's Theory of Special Relativity, then $E = hc/\lambda$ follows immediately by making using of the de Broglie equation.

You may have noticed that we have not explained why it is that the electron loses energy by emitting a photon. For the purposes of this chapter, we do not need an explanation. But something must induce the electron to leave the sanctity of its standing wave, and that 'something' is the topic of Chapter 10. For now, we are simply saying that 'in order to explain the observed patterns of light emitted by atoms it is necessary to suppose that the light is emitted when an electron drops down from one energy level to another level of lower energy'. The allowed energy levels are determined by the shape of the confining box and they vary from atom to atom because different atoms present a different environment within which their electrons are confined.

Up until now, we have made a good fist of explaining things using a very simple picture of an atom, but it isn't really good enough to pretend that electrons move around freely inside some confining box. They are moving around in the vicinity of a bunch of protons and other electrons, and to really understand atoms we must now think about how to describe this environment more accurately.

The Atomic Box

Armed with the notion of a potential, we can be more accurate in our description of atoms. Let's start with the simplest of all atoms, a hydrogen atom. A hydrogen atom is made up of just two particles: one electron and one proton. The proton is nearly 2,000 times heavier than the electron, so we can assume that it is not doing much and just sits there, creating a potential within which the electron is trapped.

The proton has a positive electric charge and the electron has an equal and opposite negative charge. As an aside, the reason why the electric charges of the proton and the electron are *exactly* equal and opposite is one of the great mysteries of physics. There is probably a very good reason, associated with some underlying theory of subatomic particles that we have yet to discover, but, as we write this book, nobody knows.

What we do know is that, because opposite charges attract, the proton is going to tug the electron towards it and, as far as pre-quantum physics is concerned, it could pull the electron inwards to arbitrarily small distances. How small would depend on the precise nature of a proton; is it a hard ball or a nebulous cloud of something? This question is irrelevant because, as we have seen, there is a minimum energy level that the electron can be in, determined (roughly speaking) by the longest wavelength quantum wave that will fit inside the potential generated by the proton. We've sketched the potential created by the proton in Figure 6.8. The deep 'hole' functions like the square well potential we met earlier except that

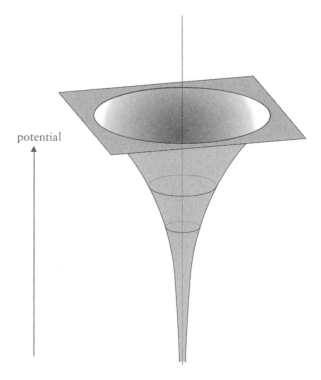

potential

Figure 6.8. The Coulomb potential well around a proton. The well is deepest where the proton is located.

the shape is not as simple. It is known as the 'Coulomb potential', because it is determined by the law describing the interaction between two electric charges, first written down by Charles-Augustin de Coulomb in 1783. The challenge is the same, however: we must find out what quantum waves can fit inside the potential, and these will determine the allowed energy levels of the hydrogen atom.

Being blunt, we might say that the way to do this is to 'solve Schrödinger's wave equation for the Coulomb potential well', which is one way to implement the clock-hopping rules. The details are technical, even for something as simple as a hydrogen atom, but fortunately we do not really learn much more than we have appreciated already. For that reason, we shall jump straight to the answer, and Figure 6.9 shows some of the resulting standing waves for an electron in a hydrogen atom. What is shown is a map of the probability to find the electron somewhere. The bright regions are where the electron is most likely to be. The real hydrogen atom is, of course, three-dimensional, and these pictures correspond to slices through the centre of the atom. The figure on the top left is the ground state wavefunction, and it tells us that the electron is, in this case, typically to be found around 1×10^{-10} m from the proton. The energies of the standing waves increase from the top left to the bottom right. The scale also changes by a factor of eight from the top left to the bottom right – in fact the bright region covering most of the top-left picture is approximately the same size as the small bright spots in the centre of the two pictures on the right. This means the electron is likely to be farther away from the proton when it is in the higher energy levels (and hence that it is more weakly bound to it). It is clear that these waves are not sine waves, which means they do not correspond to states of definite momentum. But, as we have been at pains to emphasize, they do correspond to states of definite energy.

The distinctive shape of the standing waves is due to the shape of the well and some features are worth discussing in a little more detail. The most obvious feature of the well around a proton is that it is spherically symmetric. This means that it looks the same no

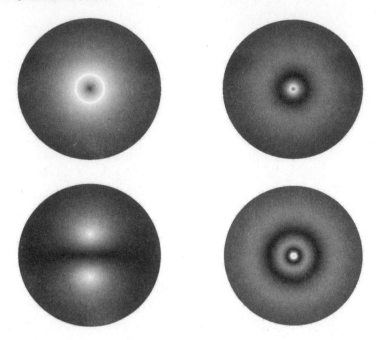

Figure 6.9. Four of the lowest energy quantum waves describing the electron in a hydrogen atom. The light regions are where the electron is most likely to be found and the proton is in the centre. The top-right and bottom-left pictures are zoomed out by a factor of 4 relative to the first and the bottom-right picture is zoomed out by a factor of 8 relative to the first. The first picture is around 3×10^{-10} m across.

matter which angle you view it from. To picture this, think of a basketball with no markings on it: it's a perfect sphere and it will look exactly the same no matter how you rotate it around. Perhaps we might dare to think of an electron inside a hydrogen atom as if it were trapped inside a tiny basketball? This is certainly more plausible than saying the electron is trapped in a square well and, remarkably, there is a similarity. Figure 6.10 shows, on the left, two of the lowest-energy standing sound waves that can be produced within a basketball. Again we have taken a slice through the ball, and the air pressure within the ball varies from black to white as the pressure increases. On the right are two possible electron standing

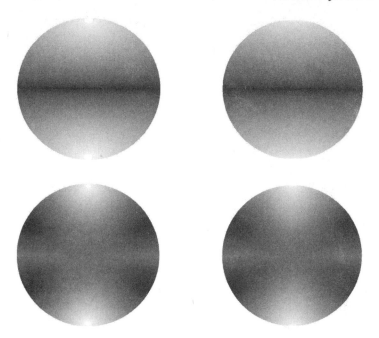

Figure 6.10: Two of the simplest standing sound waves inside a basketball (left) compared to the corresponding electron waves in a hydrogen atom (right). They are very similar. The top picture for hydrogen is a close-up of the central region in the bottom left picture in Figure 6.9.

waves in a hydrogen atom. The pictures are not identical, but they are very similar. So, it is not entirely stupid to imagine that the electron within a hydrogen atom is being trapped within something akin to a tiny basketball. This picture really serves to illustrate the wavelike behaviour of quantum particles, and it hopefully takes some of the mystery out of things: understanding the electron in a hydrogen atom is not more complicated than understanding how the air vibrates inside a basketball.

Before we leave the hydrogen atom, we would like to say a little more about the potential created by the proton and how it is that the electron can leap from a higher energy level to a lower one with the emission of a photon. We avoided any discussion of how the

proton and the electron communicate with each other, quite legitimately, by introducing the idea of a potential. This simplification
allowed us to understand the quantization of energy for trapped
particles. But if we want a serious understanding of what's going
on, we should try to explain the underlying mechanism for trapping
particles. In the case of a particle moving in an actual box, we might
imagine some impenetrable wall that is presumably made up of
atoms, and the particle is prevented from passing through the wall
by interacting with the atoms within it. A proper understanding
of 'impenetrability' comes from understanding how the particles
interact with each other. Likewise, we said that the proton in a hydrogen atom 'produces a potential' in which the electron moves, and we
said that the potential traps the electron in a manner analogous to the
way a particle is trapped in a box. That too ducks the deeper issue,
because clearly the electron interacts with the proton and it is that
interaction which dictates how the electron is confined.

In Chapter 10 we'll see that we need to supplement the quantum
rules we've articulated so far with some new rules dealing with
particle interactions. At the moment, we have very simple rules:
particles hop around, carrying imaginary clocks which wind back
by clearly specified amounts depending on the size of the hop. All
hops are allowed, and so a particle can hop from A to B via an infinity of different routes. Each route delivers its own quantum clock to
B and we must add up the clocks to determine a single resultant
clock. That clock then tells us the chance of actually finding the
particle at B. Adding interactions into the game turns out to be
surprisingly simple. We supplement the hopping rules with a new
rule, stating that a particle can emit or absorb another particle. If
there was one particle before the interaction, then there can be two
particles afterwards; if there were two particles before the interaction, then there can be one particle afterwards. Of course, if we are
going to work out the maths then we need to be more precise about
which particles can fuse together or split apart, and we need to say
what happens to the clock that each particle carries when it interacts. This is the subject of Chapter 10, but the implications for atoms

should be clear. If there is a rule saying that an electron can interact by emitting a photon, then we have the possibility that the electron in a hydrogen atom can spit out a photon, lose energy and drop down to a lower energy level. It could also absorb a photon, gain energy and leap up to a higher energy level.

The existence of spectral lines indicates that this is what is happening, and this process is ordinarily heavily biased one way. In particular, the electron can spit out a photon and lose energy at any time, but the only way it can gain energy and jump up to a higher energy level is if there is a photon (or some other source of energy) available to collide with it. In a gas of hydrogen, such photons are typically few and far between, and an atom in an excited state is much more likely to emit a photon than absorb one. The net effect is that hydrogen atoms tend to de-excite, by which we mean that emission wins over absorption and, given time, the atom will make its way down to the $n = 1$ ground state. This is not always the case, because it is possible to arrange to continually excite atoms by feeding them energy in a controlled way. This is the basis of a technology that has become ubiquitous: the laser. The basic idea of a laser is to pump energy into atoms, excite them, and collect the photons that are produced when the electrons drop down in energy. Those photons are very useful for reading data with high precision from the surface of a CD or DVD: quantum mechanics affects our lives in myriad ways.

In this chapter, we have succeeded in explaining the origin of spectral lines using the simple idea of quantized energy levels. It would seem we have a way of thinking about atoms that works. But something is not quite right. We are missing one final piece of the jigsaw, without which we have no chance of explaining the structure of atoms heavier than hydrogen. More prosaically, we will also be unable to explain why we don't fall through the floor, and that is problematic for our best theory of Nature. The insight we are looking for comes from the work of Austrian physicist Wolfgang Pauli.

7. The Universe in a Pin-head (and Why We Don't Fall Through the Floor)

That we do not fall through the floor is something of a mystery. To say the floor is 'solid' is not very helpful, not least because Rutherford discovered that atoms are almost entirely empty space. The situation is made even more puzzling because, as far as we can tell, the fundamental particles of Nature are of no size at all.

Dealing with particles 'of no size' sounds problematic, and perhaps impossible. But nothing we said in the previous chapters presupposed or required that particles have any physical extent. The notion of truly point-like objects need not be wrong, even if it flies in the face of common sense – if indeed the reader has any common sense left at this stage of a book on quantum theory. It is, of course, entirely possible that a future experiment, perhaps even the Large Hadron Collider, will reveal that electrons and quarks are not infinitesimal points, but for now this is not mandated by experiment and there is no place for 'size' in the fundamental equations of particle physics. That's not to say that point particles don't have their problems – the idea of a finite charge compressed into an infinitely small volume is a thorny one – but so far the theoretical pitfalls have been circumvented. Perhaps the outstanding problem in fundamental physics, the development of a quantum theory of gravity, hints at finite extent, but the evidence is just not there to force physicists to abandon the idea of elementary particles. To be emphatic: point-like particles are really of no size and to ask 'What happens if I split an electron in half?' makes no sense at all – there is no meaning to the idea of 'half an electron'.

A pleasing bonus of working with elementary fragments of matter that have no size at all is that we don't have any trouble with the idea that the entire visible Universe was once compressed into a volume the size of a grapefruit, or even a pin-head. Mind-boggling though that may seem – it's hard enough to imagine compressing a mountain to the size of a pea, never mind a star, a galaxy, or the 350 billion large galaxies in the observable Universe – there is absolutely no reason why this shouldn't be possible. Indeed, present-day theories of the origins of structure in the Universe deal directly with its properties when it was in such an astronomically dense state. Such theories, whilst outlandish, have a good deal of observational evidence in their favour. In the final chapter we will meet objects with densities, if not at the 'Universe in a pin-head' scale, then certainly in 'mountain in a pea' territory: white dwarves are objects with the mass of a star squashed to the size of the Earth, and neutron stars have similar masses condensed into perfect, city-sized spheres. These objects are not science fiction; astronomers have observed them and made high-precision measurements of them, and quantum theory will allow us to calculate their properties and compare them with the observational data. As a first step on the road to understanding white dwarves and neutron stars, we will need to address the more prosaic question with which we began this chapter: if the floor is largely empty space, why do we not fall through it?

This question has a long and venerable history, and the answer was not established until surprisingly recently, in 1967, in a paper by Freeman Dyson and Andrew Lenard. They embarked on the quest because a colleague had offered a bottle of vintage champagne to anyone who could prove that matter shouldn't simply collapse in on itself. Dyson referred to the proof as extraordinarily complicated, difficult and opaque, but what they showed was that matter can only be stable if electrons obey something called the Pauli Exclusion Principle, one of the most fascinating facets of our quantum universe.

We shall begin with some numerology. We saw in the last chapter that the structure of the simplest atom, hydrogen, can be understood by searching for the allowed quantum waves that fit inside the proton's potential well. This allowed us to understand, at least qualitatively, the distinctive spectrum of the light emitted from hydrogen atoms. If we had had the time, we could have calculated the energy levels in a hydrogen atom. Every undergraduate physics student performs this calculation at some stage in their studies and it works beautifully, agreeing with the experimental data. As far as the last chapter was concerned, the 'particle in a box' simplification was good enough because it contains all the key points that we wanted to highlight. However, there is a feature of the full calculation that we shall need, which comes about because the real hydrogen atom is extended in three dimensions. For our particle in a box example, we only considered one dimension and obtained a series of energy levels labelled by a single number that we called n. The lowest energy level was labelled $n = 1$, the next $n = 2$ and so on. When the calculation is extended to the full three-dimensional case it turns out, perhaps unsurprisingly, that three numbers are needed to characterize all of the allowed energy levels. These are traditionally labelled n, l and m, and they are referred to as quantum numbers (in this chapter, m is not to be confused with the mass of the particle). The quantum number n is the counterpart of the number n for a particle in a box. It takes on integer values ($n = 1$, 2, 3, etc.) and the particle energies tend to increase as n increases. The possible values of l and m turn out to be linked to n; l must be smaller than n and it can be zero, e.g. if $n = 3$ then l can be 0, 1 or 2. m can take on any value ranging from minus l to plus l in integer steps. So if $l = 2$ then m can be equal to -2, -1, 0, 1 or 2. We are not going to explain where those numbers come from, because it won't add anything to our understanding. Suffice to say that the four waves in Figure 6.9 have $(n,l) = (1,0)$, $(2,0)$, $(2,1)$ and $(3,0)$ respectively (all have $m = 0$).[1]

As we have said, the quantum number n is the main number

---

1. Technically, as we mentioned in the previous chapter, because the potential well around the proton is spherically symmetric rather than a square box, the solution to the Schrödinger equation must be proportional to a spherical harmonic. The associated angular dependence gives rise to the l and m quantum numbers. The radial dependence of the solution gives rise to the principal quantum number n.

controlling the values of the allowed energies of the electrons. There is also a small dependence of the allowed energies upon the value of l but it only shows up in very precise measurements of the emitted light. Bohr didn't consider it when he first calculated the energies of the spectral lines of hydrogen, and his original formula was expressed entirely in terms of n. There is absolutely no dependence of the electron energy upon m unless we put the hydrogen atom inside a magnetic field (in fact m is known as the 'magnetic quantum number'), but this certainly doesn't mean that it isn't important. To see why, let's get on with our bit of numerology.

If $n = 1$ then how many different energy levels are there? Applying the rules we stated above, l and m can both only be 0 if $n = 1$, and so there is just the one energy level.

Now let's do it for $n = 2$: l can take on two values, 0 and 1. If $l = 1$, then m can be equal to -1, 0 or $+1$, which is 3 more energy levels, making 4 in total.

For $n = 3$, l can be 0, 1 or 2. For $l = 2$, m can be equal to -2, -1, 0, $+1$, or $+2$, giving 5 levels. So in total, there are $1 + 3 + 5 = 9$ levels for $n = 3$. And so on.

Remember those numbers for the first three values of n: 1, 4 and 9. Now take a look at Figure 7.1, which shows the first four rows of the periodic table of the chemical elements, and count how many elements there are in each row. Divide that number by 2, and you'll get 1, 4, 4 and 9. The significance of all this will soon be revealed.

Credit for arranging the chemical elements in this way is usually

Group	1	2	3	4	5	6	7	8	9	10	11	12	13	14	15	16	17	18
1	1 H																	2 He
2	3 Li	4 Be											5 B	6 C	7 N	8 O	9 F	10 Ne
3	11 Na	12 Mg											13 Al	14 Si	15 P	16 S	17 Cl	18 Ar
4	19 K	20 Ca	21 Sc	22 Ti	23 V	24 Cr	25 Mn	26 Fe	27 Co	28 Ni	29 Cu	30 Zn	31 Ga	32 Ge	33 As	34 Se	35 Br	36 Kr

Figure 7.1. The first four rows of the periodic table.

given to the Russian chemist Dmitri Mendeleev, who presented it to the Russian Chemical Society on 6 March 1869, which was a good few years before anyone had worked out how to count the allowed energy levels in a hydrogen atom. Mendeleev arranged the elements in order of their atomic weights, which in modern language corresponds to the number of protons and neutrons inside the atomic nucleus, although of course he didn't know that at the time either. The ordering of the elements actually corresponds to the number of protons inside the nucleus (the number of neutrons is irrelevant) but for the lighter elements this makes no difference, which is why Mendeleev got it right. He chose to arrange the elements in rows and columns because he noticed that certain elements had very similar chemical properties, even though they had different atomic weights; the vertical columns group together such elements – helium, neon, argon and krypton on the far right of the table are all unreactive gases. Mendeleev didn't just get the pattern right, he also predicted the existence of new elements to fill gaps in his table: elements 31 and 32 (gallium and germanium) were discovered in 1875 and 1886. These discoveries confirmed that Mendeleev had uncovered something deep about the structure of atoms, but nobody knew what.

What is striking is that there are two elements in row one, eight in rows two and three and eighteen in row four, and those numbers are exactly twice the numbers we just worked out by counting the allowed energy levels in hydrogen. Why is this?

As we have already mentioned, the elements in the periodic table are ordered from left to right in a row by the number of protons in the nucleus, which is the same as the number of electrons they contain. Remember that all atoms are electrically neutral – the positive electric charges of the protons are exactly balanced by the negative charges of the electrons. There is clearly something interesting going on that relates the chemical properties of the elements to the allowed energies that the electrons can have when they orbit around a nucleus.

We can imagine building up heavier atoms from lighter ones by adding protons, neutrons and electrons one at a time, bearing in mind that whenever we add an extra proton into the nucleus we should add an extra electron into one of the energy levels. The exercise in numerology will generate the pattern we see in the periodic table if we simply assert that each energy level can contain two and only two electrons. Let's see how this works.

Hydrogen has only one electron, so that would slot into the $n = 1$ level. Helium has two electrons, which would both fit into the $n = 1$ level. Now the $n = 1$ level is full up. We must add a third electron to make lithium, but it will have to go into the $n = 2$ level. The next seven electrons, corresponding to the next seven elements (beryllium, boron, carbon, nitrogen, oxygen, fluorine and neon), can also sit in a level with $n = 2$ because that has four slots available, corresponding to $l = 0$ and $l = 1$, $m = -1, 0$ and $+1$. In that way we can account for all of the elements up to neon. With neon, the $n = 2$ levels are all full and we must move to $n = 3$, starting with sodium. The next eight electrons, one by one, start to fill up the $n = 3$ levels; first the electrons go into $l = 0$, and then into $l = 1$. That accounts for all the elements in the third row, up to argon. The fourth row of the table can be explained if we assume that it contains all of the remaining $n = 3$ electrons (i.e. the ten electrons with $l = 2$) and the $n = 4$ electrons with $l = 0$ and 1 (which makes eight electrons), making the magic number of eighteen electrons in total. We've sketched how the electrons fill up the energy levels for the heaviest element in our table, krypton (which has thirty-six electrons) in Figure 7.2.

To elevate all of what we just said to science rather than numerology we have some explaining to do. Firstly, we need to explain why the chemical properties are similar for elements in the same vertical column. What is clear from our scheme is that the first element in each of the first three rows starts off the process of filling levels with increasing values of n. Specifically, hydrogen starts things off with a single electron in the otherwise empty $n = 1$ level, lithium starts off the second row with a single electron in the $n = 2$ level and sodium starts the third row with a single electron in the otherwise empty $n = 3$ level. The third row is a little odd because the $n = 3$ level can

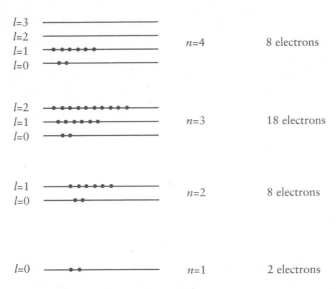

Figure 7.2. Filling the energy levels of krypton. The dots represent electrons and the horizontal lines represent the energy levels, labelled by the quantum numbers n, l and m. We have grouped together levels with different values of m but the same values of n and l.

hold eighteen electrons and there are not eighteen elements in the third row. We can guess at what is happening though – the first eight electrons fill up the $n = 3$ levels with $l = 0$ and $l = 1$, and then (for some reason) we should switch to the fourth row. The fourth row now contains the remaining ten electrons from the $n = 3$ levels with $l = 2$ and the eight electrons from the $n = 4$ levels with $l = 0$ and $l = 1$. The fact that the rows are not entirely correlated with the value of n indicates that the link between the chemistry and the energy-level counting is not as simple as we have been making out. However, it is now known that potassium and calcium, the first two elements in the fourth row, do have electrons in the $n = 4$, $l = 0$ level and that the next ten elements (from scandium to zinc) have their electrons in the belated $n = 3$, $l = 2$ levels.

To understand why the filling up of the $n = 3$ and $l = 2$ levels is deferred until after calcium requires an explanation of why the $n = 4$, $l = 0$ levels, which contain the electrons in potassium and calcium, is of lower energy than the $n = 3$, $l = 2$ levels. Remember, the 'ground state' of an atom will be characterized by the lowest-energy configuration of the electrons, because any excited state can always lower its energy by the emission of a photon. So when we have been saying that 'this atom contains these electrons sitting in those energy levels' we are telling you the lowest energy configuration of the electrons. Of course, we have not made any attempt to actually compute the energy levels, so we aren't really in a position to rank them in order of energy. In fact it is a very difficult business to calculate the allowed electron energies in atoms with more than two electrons, and even the two-electron case (helium) is not so easy. The simple idea that the levels are ranked in order of increasing n comes from the much easier calculation for the hydrogen atom, where it is true that the $n = 1$ level has the lowest energy followed by the $n = 2$ levels, then come the $n = 3$ levels and so on.

The obvious implication of what we just said is that the elements on the far right of the periodic table correspond to atoms in which a set of levels has just been completely filled. In particular, for helium the $n = 1$ level is full, whilst for neon the $n = 2$ level is full, and for argon the $n = 3$ level is fully populated, at least for $l = 0$ and $l = 1$. We can develop these ideas a little further and understand some important ideas in chemistry. Fortunately we aren't writing a chemistry textbook, so we can be brief and, at the risk of dismissing an entire subject in a single paragraph, here we go.

The key observation is that atoms can stick together by sharing electrons – we will meet this idea in the next chapter when we explore how a pair of hydrogen atoms can bind to make a hydrogen molecule. The general rule is that elements 'like' to have all their energy levels neatly filled up. In the case of helium, neon, argon and krypton, the levels are already completely full, and so they are 'happy' on their own – they don't 'bother' reacting with anything. For the other elements, they can 'try' to fill their levels by sharing electrons with other elements. Hydrogen, for example, needs one extra electron to fill its $n = 1$ level. It can achieve this by sharing an electron with another hydrogen atom. In so doing, it forms a hydrogen molecule, with chemical symbol H_2. This is the common form in which hydrogen gas exists. Carbon has four electrons out of a possible eight in its $n = 2$, $l = 0$ and $l = 1$ levels, and would 'like' another four if possible to fill them up. It can achieve this by binding together with four hydrogen atoms to form CH_4, the gas known as methane. It can also do it by binding with two oxygen atoms, which themselves need two electrons to complete their $n = 2$ set. This leads to CO_2 – carbon dioxide. Oxygen could also complete its set by binding with two hydrogen atoms to make H_2O – water. And so on. This is the basis of chemistry: it is energetically favourable for atoms to fill their energy levels with electrons, even if that is achieved by sharing with a neighbour. Their 'desire' to do this, which ultimately stems from the principle that things tend to their lowest energy state, is what drives the formation of everything from water to DNA. In a world abundant in hydrogen, oxygen and carbon we now understand why carbon dioxide, water and methane are so common.

This is very encouraging, but we have a final piece of the jigsaw to explain: why is it that only two electrons can occupy each available energy level? This is a statement of the Pauli Exclusion Principle, and it is clearly necessary if everything we have been discussing is to hang together. Without it, the electrons would crowd together in the lowest possible energy level around every nucleus, and there would be no chemistry, which is worse than it sounds, because there would be no molecules and therefore no life in the Universe.

The idea that two and only two electrons can occupy each energy level does seem quite arbitrary, and historically nobody had any idea why it should be the case when the idea was first proposed. The initial breakthrough was made by Edmund Stoner, the son of a professional cricketer (who took eight wickets against South Africa in 1907, for those who read their *Wisden Cricketers' Almanack*) and a former student of Rutherford's who later ran the physics department at the University of Leeds. In October 1924, Stoner proposed that there should be two electrons allowed in each (n, l, m) energy level. Pauli developed Stoner's proposal and in 1925 he published a rule that Dirac named after him a year later. The Exclusion Principle, as first proposed by Pauli, states that no two electrons in an atom can share the same quantum numbers. The problem he faced was that it appeared that two electrons *could* share each set of n, l and m values. Pauli got round the problem by simply introducing a new quantum number. This was an ansatz; he didn't know what it represented, but it had to take on one of only two values. Pauli wrote that, 'We cannot give a more precise reason for this rule.' Further insight came in 1925, in a paper by George Uhlenbeck and Samuel Goudsmit. Motivated by precise measurements of atomic spectra, they identified Pauli's extra quantum number with a real, physical property of the electron known as 'spin'.

The basic idea of spin is quite simple, and dates back to 1903, well before quantum theory. Just a few years after its discovery, German physicist Max Abraham proposed that the electron was a tiny, spinning electrically charged sphere. If this were true, then electrons would be affected by magnetic fields, depending on the orientation of the field relative to their spin axis. In their 1925 paper, which was published three years after Abraham's death, Uhlenbeck and Goudsmit noted that the spinning ball model couldn't work because, in order to explain the observed data, the electron would have to be spinning faster than the speed of light. But the spirit of the idea was correct – the electron does possess a property called spin, and it does affect its behaviour in a magnetic field. Its true origin, however, is a direct and rather subtle consequence of Einstein's Theory of Special Relativity that was only properly appreciated when Paul Dirac wrote down an equation describing the quantum behaviour of the electron in 1928. For our purposes, we shall need only acknowledge that electrons do come in two types, which we refer to as 'spin up' and 'spin down', and the two are distinguished by having opposite values of their angular momentum, i.e. it is like they are spinning in opposite directions. It's a pity that Abraham died just a few years before the true nature of electron spin was discovered, because he never gave up his conviction that the electron was a little sphere. In his obituary in 1923, Max Born and Max Von Laue wrote: 'He was an honourable opponent who fought with honest weapons and who did not cover up a defeat by lamentation and nonfactual arguments . . . He loved his absolute ether, his field equations, his rigid electron, just as a youth loves his first flame, whose memory no later experience can extinguish.' If only all of one's opponents were like Abraham.

Our goal in the remainder of this chapter is to explain why it is that electrons behave in the strange way articulated by the Exclusion Principle. As ever, we shall make good use of those quantum clocks.

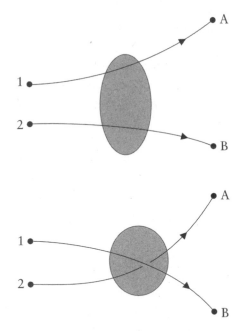

Figure 7.3. Two electrons scattering.

We can attack the question by thinking about what happens when two electrons 'bounce' off each other. Figure 7.3 illustrates a particular scenario where two electrons, labelled '1' and '2', start out somewhere and end up somewhere else. We have labelled the final locations A and B. The shaded blobs are there to remind us that we have not yet thought about just what happens when two electrons interact with each other (the details are irrelevant for the purposes of this discussion). All we need to imagine is that electron 1 hops from its starting place and ends up at the point labelled A. Likewise, electron 2 ends up at the point labelled B. This is what is illustrated in the top of the two pictures in the figure. In fact, the argument we are about to present works fine even if we ignore the possibility that the electrons might interact. In that case, electron 1 hops to A oblivious to the meanderings of electron 2 and the probability of finding electron 1 at A and electron 2 at B would be simply a product of two independent probabilities.

For example, suppose the probability of electron 1 hopping to point A is 45% and the probability of electron 2 hopping to point B is 20%. The probability of finding electron 1 at A *and* electron 2 at B is $0.45 \times 0.2 = 0.09 = 9\%$. All we are doing here is using the logic that says that the chances of tossing a coin and getting 'tails' *and* rolling a dice and getting a 'six' at the same time is one-half multiplied by one-sixth, which is equal to $\frac{1}{12}$ (i.e. just over 8%).[2]

As the figure illustrates, there is a second way that the two electrons can end up at A and B. It is possible for electron 1 to hop to B whilst electron 2 ends up at A. Suppose that the chance of finding electron 1 at B is 5% and the chance of finding electron 2 at A is 20%. Then the probability of finding electron 1 at B *and* electron 2 at A is $0.05 \times 0.2 = 0.01 = 1\%$.

We therefore have two ways of getting our two electrons to A and B – one with a probability of 9% and one with a probability of 1%. The probability of getting one electron at A and one at B, if we don't care which is which, should therefore be $9\% + 1\% = 10\%$. Simple; but wrong.

The error is in supposing that it is possible to say which electron arrives at A and which one arrives at B. What if the electrons are

---

2. We will learn in Chapter 10 that accounting for the possibility that the two electrons interact with each other means we need to calculate the probability to find electron 1 at A and electron 2 at B 'all at once' because it does not reduce to a multiplication of two independent probabilities. But that is a detail as far as this chapter is concerned.

identical to each other in every way? This might sound like an irrelevant question, but it isn't. Incidentally, the suggestion that quantum particles might be strictly identical was first made in relation to Planck's black body radiation law. A little-known physicist called Ladislas Natanson had pointed out, as far back as 1911, that Planck's law was incompatible with the assumption that photons could be treated as identifiable particles. In other words, if you could tag a photon and track its movements, then you wouldn't get Planck's law.

If electrons 1 and 2 are absolutely identical then we must describe the scattering process as follows: initially there are two electrons, and a little later there are still two electrons located in different places. As we've learnt, quantum particles do not travel along well-defined trajectories, and this means that there really is no way of tracking them, even in principle. It therefore makes no sense to claim electron 1 appeared at A and electron 2 at B. We simply can't tell, and it is therefore meaningless to label them. This is what it means for two particles to be 'identical' in quantum theory. Where does this line of reasoning take us?

Look at the figure again. For this particular process, the two probabilities we associated with the two diagrams (9% and 1%) are not wrong. They are, however, not the whole story. We know that quantum particles are described by clocks, so we should associate a clock with electron 1 arriving at A with a size equal to the square root of 45%. Likewise there is a clock associated with electron 2 arriving at B and it has a size equal to the square root of 20%.

Now comes a new quantum rule – it says that we are to associate a single clock with the process as a whole, i.e. there is a clock whose size squared is equal to the probability to find electron 1 at A and electron 2 at B. In other words, there is a single clock associated with the upper picture in Figure 7.3. We can see that this clock must have a size equal to the square root of 9%, because that is the probability for the process to happen. But what time does it read? Answering this question is the domain of Chapter 10 and it involves the idea of clock multiplication. As far as this chapter is concerned, we don't need to know the time, we only need the important new rule that we have just stated, but which is worth repeating because it is a very general statement in quantum theory: we should associate a *single* clock with each possible way that an *entire process* can happen. The clock we associate with finding a single particle at a single location is the simplest illustration of this rule, and we have managed to get this far in the book with it. But it is a special case, and as soon as we start to think about more than one particle we need to extend the rule.

This means that there is a clock of size equal to 0.3 associated with the upper picture in the figure. Likewise, there is a second clock of size equal to 0.1 (because 0.1 squared is $0.01 = 1\%$) associated with the lower picture in the figure. We therefore have two clocks and we want a way to use them to determine the probability to find an electron at A and another at B. If the two electrons were distinguishable then the answer would be simple – we would just add together the probabilities (and not the clocks) associated with each possibility. We would then obtain the answer of 10%.

But if there is absolutely no way of telling which of the diagrams actually happened, which is the case if the electrons are indistinguishable from each other, then following the logic we've developed for a single particle as it hops from place to place, we should seek to combine the clocks. What we are after is a generalization of the rule which states that, for one particle, we should add together the clocks associated with all of the different ways that the particle can reach a particular point in order to determine the probability to find the particle at that point. For a system of many identical particles, we should combine together all the clocks associated with all of the different ways that the particles can reach a set of locations in order to determine the probability that particles will be found at those locations. This is important enough to merit reading a few times – it should be clear that this new law for combining clocks is a direct generalization of the rule we have been using for a single particle. You may have noticed that we have been very careful with our wording, however. We did not say that the clocks should necessarily be *added* together – we said that they should be *combined* together. There is a good reason for our caution.

The obvious thing to do would be to add the clocks together. But before leaping in we should ask whether there is a good reason why this is correct. This is a nice example of not taking things for granted in physics – exploring our assumptions often leads to new insights, as it will do in this instance. Let's take a step back, and think of the most general thing we could imagine. This would be to allow for the possibility of giving one of the clocks a turn or a shrink (or expansion) before we add them. Let's explore this possibility in more detail.

What we are doing is saying, 'I have two clocks and I want to combine them to make a single clock, so that I can use that to tell me what the probability is for the two electrons to be found at A and B. How should I combine them?' We are not pre-empting the answer, because we want to understand if adding clocks together really is the rule we should use. It turns out that we do not have much freedom at all, and simply adding clocks is, intriguingly, one of only *two* possibilities.

To streamline the discussion, let's refer to the clock corresponding to particle 1 hopping to A *and* particle 2 hopping to B as clock 1. This is the clock associated with the upper picture in Figure 7.3. Clock 2 corresponds to the other option, where particle 1 hops to B instead. Here is an important realization: if we give clock 1 a turn before adding it to clock 2, then the final probability we calculate must be the same as if we choose to give clock 2 the same turn before adding it to clock 1.

To see this, notice that swapping the labels A and B around in our diagrams clearly cannot change anything. It is just a different way of describing the same process. But swapping A and B around swaps the diagrams in Figure 7.3 around too. This means that if we decide to wind clock 1 (corresponding to the upper picture) before adding it to clock 2, then this must correspond precisely to winding the clock 2 before adding it to clock 1, after we've swapped labels. This piece of logic is crucial, so it's worth hammering home. Because we have assumed that there is no way of telling the difference between the two particles, then we are allowed to swap the labels around.

This implies that a turn on clock 1 *must* give the same answer as when we apply the same turn to clock 2, because there is no way of telling the clocks apart.

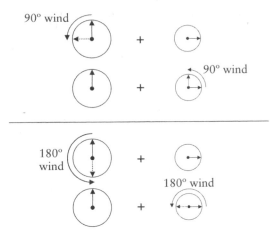

Figure 7.4. The upper part of the figure illustrates that adding clocks 1 and 2 together after winding clock 1 by 90 degrees is not the same as adding them together after winding clock 2 by 90 degrees. The lower part illustrates the interesting possibility that we could wind one of the clocks by 180 degrees before adding.

This is not a benign observation – it has a very important consequence, because there are only *two* possible ways of playing around with the winding and shrinking of clocks before adding them together that will deliver a final clock with the property that it does not depend upon which of the original clocks gets the treatment.

This is illustrated in Figure 7.4. The top half of the figure illustrates that, if we wind clock 1 by 90 degrees and add it to clock 2 then the resultant clock is not of the same size as the resultant we would get if we instead wound clock 2 by 90 degrees and add it to clock 1. We can see this because, if we first wind clock 1, the new hand, represented by the dotted arrow, points in the opposite direction to clock 2's hand, and therefore partly cancels it out. Winding clock 2 instead leaves its hand pointing in the same direction as clock 1's, and now the hands will add together to form a larger hand.

It should be clear that 90 degrees is not special, and that other angles will also give resultant clocks that depend upon which of clocks 1 and 2 we decided to wind.

The obvious exception is a clock wind of zero degrees, because winding clock 1 by zero degrees before adding to clock 2 is obviously exactly the same as winding clock 2 by zero degrees before adding to clock 1. This means that adding clocks together without any wind is a viable possibility. Similarly, winding both clocks by the same amount would work, but that really is just the same as the 'no winding' situation and corresponds simply to redefining what we call '12 o'clock'. This is tantamount to saying that we are always free to wind every clock around by some amount, as long as we do that to every clock. This will never impact on the probabilities we are trying to compute.

The lower part of Figure 7.4 illustrates that there is, perhaps surprisingly, one other way we can combine the clocks: we could turn one of them through 180 degrees before adding them together. This does not produce exactly the same clock in the two cases but it does produce the same size of clock, and that means it leads to the same probability to find one electron at A and a second at B.

A similar line of reasoning rules out the possibility of shrinking or expanding one of the clocks before adding, because if we shrink clock 1 by some fraction before adding to clock 2 then this will not usually be the same as shrinking clock 2 by that same amount before adding it to clock 1, and there are no exceptions to that rule.

So, we have an interesting conclusion to draw. Even though we started out by allowing ourselves complete freedom, we have discovered that, because there is no way of telling the particles apart, there are in fact only two ways we can combine the clocks: we can either add them or we can add them after first winding one or the other by 180 degrees. The truly delightful thing is that Nature exploits both possibilities.

For electrons, we have to incorporate the extra twist before adding the clocks. For particles like photons, or Higgs bosons, we have to add clocks without the twist. And so it is that Nature's particles come in two types: those which need the twist are called fermions and those without the twist are called bosons. What determines whether a particular particle is a fermion or a boson? It is the spin.

The spin is, as the name suggests, a measure of the angular momentum of a particle and it is a matter of fact that fermions always have a spin equal to some half-integer value[3] while bosons always have integer spin. We say that the electron has spin-half, the photon has spin-one and the Higgs boson has spin-zero. We have been avoiding dealing with the details of spin in this book, because it is a technical detail most of the time. However, we did need the result that electrons can come in two types, corresponding to the two possible values of their angular momentum (spin up and spin down), when we were discussing the periodic table. This is an example of a general rule that says particles of spin s generally come in $2s + 1$ types, e.g. spin $\frac{1}{2}$ particles (like electrons) come in two types, spin 1 particles come in three types and spin 0 particles come in one type. The relationship between the angular momentum of a particle and the way we are to combine clocks is known as the spin-statistics theorem, and it emerges when quantum theory is formulated so that it is consistent with Einstein's Theory of Special Relativity. More specifically, it is a direct result of making sure that the law of cause and effect is not violated. Unfortunately, deriving the spin-statistics theorem is beyond the level of this book – actually it is beyond the level of many books. In *The Feynman Lectures on Physics*, Richard Feynman has this to say:

We apologise for the fact that we cannot give you an elementary explanation. An explanation has been worked out by Pauli from complicated arguments of quantum field theory and relativity. He has shown that the two must necessarily go together, but we have not been able to find a way of reproducing his arguments at an elementary level. It appears to be one of the few places in physics where there is a rule which can be stated very simply, but for which no one has found a simple and easy explanation.

3. In units of Planck's constant divided by 2π.

Bearing in mind that Richard Feynman wrote this in a university-level textbook, we must hold up our hands and concur. But the rule is simple, and you must take our word for it that it can be proved: for fermions, you have to give a twist, and for bosons you don't. It turns out that the twist is the reason for the Exclusion Principle, and therefore for the structure of atoms; and, after all our hard work, this is now something that we can explain very simply.

Imagine moving points A and B in Figure 7.3 closer and closer together. When they are very close together, clock 1 and clock 2 must be of nearly the same size and read nearly the same time. When A and B are right on top of each other then the clocks must be identical. That should be obvious, because clock 1 corresponds to particle 1 ending up at point A and clock 2 is, in this special case, representing exactly the same thing because points A and B are on top of each other. Nevertheless, we do still have two clocks, and we must still add them together. But here is the catch: for fermions, we must give one of the clocks a twist, winding it first by 180 degrees. This means that the clocks will always read exactly 'opposite' times when A and B are in the same place – if one reads 12 o'clock then the other will read 6 o'clock – so adding them together always gives a resultant clock of zero size. That is a fascinating result, because it means that there is always zero chance of finding the two electrons at the same place: the laws of quantum physics are causing them to avoid each other. The closer they get to each other, the smaller the resultant clock, and the less likely that is to happen. This is one way to articulate Pauli's famous principle: electrons avoid each other.

Originally, we set out to demonstrate that no two identical electrons can be in the same energy level in a hydrogen atom. We have not quite shown this to be true yet, but the notion that electrons avoid each other clearly has implications for atoms and for why we do not fall through the floor. Now we can see that not only do the electrons in the atoms in our shoes push against the electrons in the floor because like-charges repel; they also push against each other because they naturally avoid each other, according to the Pauli Exclusion Principle. It turns out that, as Dyson and Lenard proved, it is the electron avoidance that really keeps us from falling through the floor, and it also forces the electrons to occupy the different energy levels inside atoms, giving them a structure, and ultimately leading to the variety of chemical elements we see in Nature. This is clearly a piece of physics with very significant consequences for everyday life. In the final chapter of this book, we will show how Pauli's principle also plays a crucial role in preventing some stars from collapsing under the influence of their own gravity.

To finish, we should explain how it is that, if no two electrons can be at the same place at the same time, then it also follows that no two electrons in an atom can have the same quantum numbers, which means that they cannot have the same energy and spin. If we consider two electrons of the same spin, then we want to show that they cannot be in the same energy level. If they were in the same energy level then necessarily each electron would be described by exactly the same array of clocks distributed through space (corresponding to the relevant standing wave). For each pair of points in space – let's denote them X and Y – there are then two clocks. Clock 1 corresponds to 'electron 1 at X' and 'electron 2 at Y', whilst clock 2 corresponds to 'electron 1 at Y' and 'electron 2 at X'. We know from our previous deliberations that these clocks should be added together after winding one of them by 6 hours in order to deduce the probability to find one electron at X and a second at Y. But if the two electrons have the same energies, then clocks 1 and 2 must be identical to each other before the crucial extra wind. After the wind, they will read 'opposite' times and, as before, add together to make a clock of no size. That happens for any particular locations X and Y, and so there is absolutely zero chance of ever finding a pair of electrons in the same standing wave configuration, and therefore with the same energy. That, ultimately, is responsible for the stability of the atoms in your body.

8. Interconnected

So far we have been paying close attention to the quantum physics of isolated particles and atoms. We have learnt that electrons sit inside atoms in states of definite energy, known as stationary states, although the atom may be in a superposition of different such states. We have also learnt that it is possible for an electron to make a transition from one energy state to another with the concurrent emission of a photon. The emission of photons in this way makes tangible the energy states in an atom; we see the characteristic colours of atomic transitions everywhere. Our physical experience, though, is of vast assemblies of atoms stuck together in clumps, and for that reason alone it is time to start pondering what happens when we stick atoms together.

The contemplation of atomic clusters is going to lead us along a road that will take in chemical bonding, the differences between conductors and insulators and, eventually, to semiconductors. These interesting materials have properties that can be exploited to build tiny devices capable of carrying out operations in basic logic. They are known as transistors, and by stringing many millions of them together we can build microchips. As we shall see, the theory of transistors is deeply quantum. It is difficult to see how they could have been invented and exploited without quantum theory, and difficult to imagine the modern world without them. They are a prime example of serendipity in science; the curiosity-led exploration of Nature that we've spent so much time describing in all its counterintuitive detail, eventually led to a revolution in our everyday lives. The dangers in trying to classify and control scientific research is beautifully summarized in the words of William Shockley, one of the inventors of the transistor and head of the Solid State Physics Group at Bell Telephone Laboratories:[1]

1. This is an excerpt from his 1956 Nobel Prize-winner's speech.

I would like to express some viewpoints about words often used to classify types of research in physics; for example, pure, applied, unrestricted, fundamental, basic, academic, industrial, practical, etc. It seems to me that all too frequently some of these words are used in a derogatory sense, on the one hand to belittle the practical objectives of producing something useful and, on the other hand, to brush off the possible long-range value of explorations into new areas where a useful outcome cannot be foreseen. Frequently, I have been asked if an experiment I have planned is pure or applied research; to me it is more important to know if the experiment will yield new and probably enduring knowledge about nature. If it is likely to yield such knowledge, it is, in my opinion, good fundamental research; and this is much more important than whether the motivation is purely esthetic satisfaction on the part of the experimenter on the one hand or the improvement of the stability of a high-power transistor on the other. It will take both types to confer the greatest benefit on mankind.

Since that comes from the inventor of perhaps the most useful device since the invention of the wheel, policy-makers and managers throughout the world would do well to pay attention. Quantum theory changed the world, and whatever new theories emerge from the cutting-edge physics being done today, they will almost certainly change our lives again.

As ever, we'll start at the beginning and extend our study of a universe containing just one particle to a universe of two. Imagine, in particular, a simple universe containing two isolated hydrogen atoms; two electrons bound in orbit around two protons that are very far apart. In a few pages we are going to start bringing the two atoms closer together to see what happens, but for now we are to suppose that they are very distant from each other.

The Pauli Exclusion Principle says that the two electrons cannot be in the same quantum state, because electrons are indistinguishable fermions. You might at first be tempted to say that, if the atoms are far apart, then the two electrons must be in very different quantum states and there is not much more to be said on the matter. But things are vastly more interesting than that. Imagine putting electron number 1 in atom number 1 and electron number 2 in atom number 2. After waiting a while it doesn't make sense to say that 'electron number 1 is still in atom number 1'. It might be in atom number 2 now because there is always the chance that the electron did a quantum hop. Remember, everything that can happen does happen, and electrons are free to roam the Universe from one instant to the next. In the language of little clocks, even if we started out with clocks describing one of the electrons clustered only in the vicinity of one of the protons, we would be forced to introduce clocks in the vicinity of the other proton at the next instant. And even if the orgy of quantum interference meant that the clocks near the other proton are very tiny, they would not be of zero size, and there would always be a finite probability that the electron could be there. The way to think more clearly about the implications of the Exclusion Principle is to stop thinking in terms of two isolated atoms and think instead of the system as a whole: we have two protons and two electrons and our task is to understand how they organize themselves. Let us simplify the situation by neglecting the electromagnetic interaction between the two electrons – this won't be a bad approximation if the protons are far apart, and it doesn't affect our argument in any important way.

What do we know about the allowed energies for the electrons in the two atoms? We don't need to do a calculation to get a rough idea; we can use what we know already. For protons that are far apart (imagine they are many miles apart), the lowest allowed energies for the electrons must surely correspond to the situation where they are bound to the protons to make two isolated hydrogen atoms. In this case, we might be tempted to conclude that the lowest energy state for the entire two-proton, two-electron system would correspond to two hydrogen atoms sitting in their lowest energy states, ignoring each other completely. But although this sounds right, it cannot be right. We must think of the system as a whole, and just like an isolated hydrogen atom, this four-particle system must have its own unique spectrum of allowed electron energies. And because of the Pauli principle, the electrons cannot both be in exactly the same energy level around each proton, blissfully ignorant of the existence each other.[2]

It seems that we must conclude that the pair of identical electrons in two distant hydrogen atoms cannot have the same energy but we have also said that we expect the electrons to be in the lowest energy level corresponding to an idealized, perfectly isolated hydrogen atom. Both those things cannot be true and a little thought indicates that the way out of the problem is for there to be not one but *two* energy levels for each level in an idealized, isolated hydrogen atom. That way we can accommodate the two electrons without violating the Exclusion Principle. The difference in the two energies must be very small indeed for atoms that are far apart, so that we can pretend the atoms are oblivious to each other. But really, they are not oblivious, because of the tendril-like reaches of the Pauli principle: if one of the two electrons is in one energy state then the other must be in the second, different energy state and this intimate link between the two atoms persists regardless of how far apart they are.

This logic extends to more than two atoms – if there are twenty-four hydrogen atoms scattered far apart across the Universe, then for every energy state in a single-atom universe there are now twenty-four energy states, all taking on almost but not quite the same values. When an electron in one of the atoms settles into a particular state it does so in full 'knowledge' of the states of each of the other twenty-three electrons, regardless of their distance away. And so, every electron in the Universe knows about the state of every other electron. We need not stop there – protons and neutrons are fermions too, and so every proton knows about every other proton and every neutron knows about every other neutron. There is an intimacy between the particles that make up our Universe that extends

2. For the sake of this discussion we are ignoring the electron's spin. What we have said still applies if we imagine that it refers to two electrons of the same spin.

across the entire Universe. It is ephemeral in the sense that for particles that are far apart the different energies are so close to each other as to make no discernible difference to our daily lives.

This is one of the weirdest-sounding conclusions we've been led to so far in the book. Saying that every atom in the Universe is connected to every other atom might seem like an orifice through which all sorts of holistic drivel can seep. But there is nothing here that we haven't met before. Think about the square well potential we thought about in Chapter 6. The width of the well determines the allowed spectrum of energy levels, and as the size of the well is changed, the energy level spectrum changes. The same is true here in that the shape of the well inside which our electrons are sitting, and therefore the energy levels they are allowed to occupy, is determined by the positions of the protons. If there are two protons, the energy spectrum is determined by the position of both of them. And if there are 10^{80} protons forming a universe, then the position of every one of them affects the shape of the well within which 10^{80} electrons are sitting. There is only ever one set of energy levels and when anything changes (e.g. an electron changes from one energy level to another) then everything else must instantaneously adjust itself such that no two fermions are ever in the same energy level.

The idea that the electrons 'know' about each other instantaneously sounds like it has the potential to violate Einstein's Theory of Relativity. Perhaps we can build some sort of signalling apparatus that exploits this instantaneous communication to transmit information at faster-than-light speeds. This apparently paradoxical feature of quantum theory was first appreciated in 1935, by Einstein in collaboration with Boris Podolsky and Nathan Rosen; Einstein called it 'spooky action at a distance' and did not like it. It took some time before people realized that, despite its spookiness, it is impossible to exploit these long-range correlations to transfer information faster than the speed of light and that means the law of cause and effect can rest safe.

This decadent multiplicity of energy levels is not just an esoteric device to evade the constraints of the Exclusion Principle. In fact, it is anything but esoteric because this is the physics behind chemical bonding. It is also the key idea in explaining why some materials conduct electricity whilst others do not and, without it, we would not understand how a transistor works. To begin our journey to the transistor, we are going to go back to the simplified 'atom' we met in Chapter 6, when we trapped an electron inside a potential well. To be sure, this simple model didn't allow us to compute the correct spectrum of energies in a hydrogen atom, but it did teach us about the behaviour of a single atom and it will serve us well here too. We are going to use two square wells joined together to make a toy model of two adjacent hydrogen atoms. We'll think first about the case where there is a single electron moving in the potential created by two protons. The upper picture in Figure 8.1 illustrates how we'll do it. The potential is flat except where it dips down to make two wells, which mimic the influence of the two protons in their ability to trap electrons. The step in the middle helps keep the electron trapped either on the left or on the right, provided it is high enough. In the technical parlance, we say that the electron is moving in a double-well potential.

Our first challenge is to use this toy model to understand what happens when we bring two hydrogen atoms together – we will see that when they get close enough they bind together, to make a molecule. After that, we shall contemplate more than two atoms and that will allow us to appreciate what happens inside solid matter.

If the wells are very deep, we can use the results from Chapter 6 to determine what the lowest-lying energy states should correspond to. For a single electron in a single square well, the lowest energy state is described by a sine wave of wavelength equal to twice the size of the box. The next-to-lowest energy state is a sine wave whose wavelength is equal to the size of the box, and so on. If we put an electron into one side of a double-well, and if the well is deep enough, then the allowed energies must be close to those for an electron trapped in a single deep well, and its wavefunction should

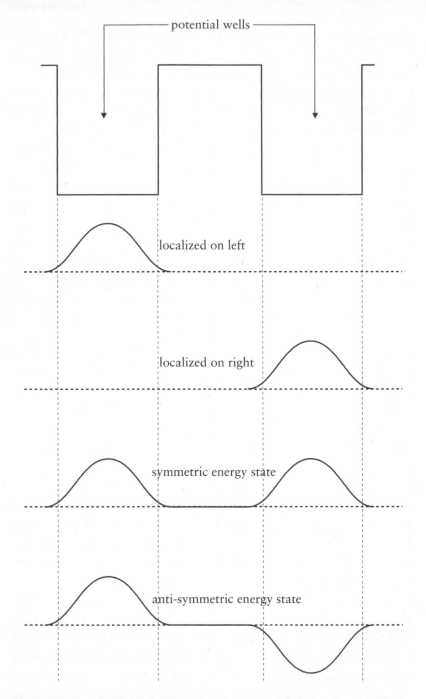

Figure 8.1. The double-well potential at the top and, below it, four interesting wavefunctions describing an electron in the potential. Only the bottom two correspond to an electron of definite energy.

therefore look quite like a sine wave. It is to the small differences between a perfectly isolated hydrogen atom and a hydrogen atom in a distantly separated pair to which we now turn our attention.

We can safely anticipate that the top two wavefunctions drawn in Figure 8.1 correspond to those for a single electron when it is located either in the left well or the right well (remember we use 'well' and 'atom' interchangeably). The waves are approximately sine waves, with a wavelength equal to twice the width of the well. Because the wavefunctions are identical in shape we might say that they should correspond to particles with equal energies. But this can't be right because, as we have already said, there must be a tiny probability that, no matter how deep the wells or how widely separated they are, the electron can hop from one to the other. We have hinted at this by sketching the sine waves as 'leaking' slightly through the walls of the well, representing the fact that there is a very small probability to find non-zero clocks in the adjacent well.

The fact that the electron always has a finite probability of leaping from one well to the other means that the top two wavefunctions in Figure 8.1 cannot possibly correspond to an electron of definite energy, because we know from Chapter 6 that such an electron is described by a standing wave whose shape does not change with time or, equivalently, a bunch of clocks whose sizes never change with time. If, as time advances, new clocks are spawned in the originally empty well then the shape of the wavefunction will most certainly change with time. What then, does a state of definite energy look like for a double well? The answer is that the states must be more democratic, and express an equal preference to find the electron in either well. This is the only way to make a standing wave and stop the wavefunction sloshing back and forth from one well to the other.

The lower two wavefunctions we've sketched in Figure 8.1 have this property. These are what the lowest-lying energy states actually look like. They are the only possible stationary states we can build that look like the 'single-well' wavefunctions in each individual well, and also describe an electron that is equally likely to be found in either well. They are in fact the *two* energy states that we deduced must be present if we are to put two electrons into orbit around two distant protons to make two almost identical hydrogen atoms in a way consistent with the Pauli principle. If one electron is described by one of these two wavefunctions, then the other electron must be described by the other – this is what the Pauli principle demands.[3] For deep enough wells, or if the atoms are far enough apart, the two energies will be almost equal, and almost equal to the lowest energy of a particle trapped in a single isolated well. We should not worry that one of the wavefunctions looks partly upside-down – remember it is only the size of a clock that matters when determining the probability to find the particle at some place. In other words, we could invert all the wavefunctions we've drawn in this book and not change the physical content of anything at all. The 'partly upside-down' wavefunction (labelled 'anti-symmetric energy state' in the figure) therefore still describes an equal superposition of an electron trapped in the left-hand well and an electron trapped in the right-hand well. Crucially though, the symmetric and anti-symmetric wavefunctions are not *exactly* the same (they could not be, otherwise Pauli would be upset). To see this, we need to look at the behaviour of these two lowest-energy wavefunctions in the region *between* the wells.

One wavefunction is symmetric about the centre of the two wells, and the other is anti-symmetric (they are labelled as such in the figure). By 'symmetric' we mean that the wave on the left is the mirror image of the wave on the right. For the 'antisymmetric' wave, the wave on the left is the mirror image of the wave on the right only after it has been turned upside down. The terminology is not very important, but what does matter is that the two waves are different in the region between the two wells. It is this small difference that means that they describe states with very slightly different energies. In fact, the symmetric wave is the one with the lower energy. So turning one of the waves upside down does in fact mat-

3. Recall we have in mind two identical electrons, i.e. they have equal spin.

ter, but only a tiny amount if the wells are deep enough or far enough apart.

It can certainly be confusing to think in terms of particles with definite energy because, as we have just seen, they are described by wavefunctions that are of equal size in either well. This genuinely does mean that there is an equal chance of finding the electron in either well when we look for it, even if the wells are separated by an entire Universe.

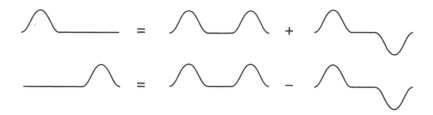

Figure 8.2. Upper: an electron localized in the left well can be understood as the *sum* of the two lowest energy states. Lower: likewise, an electron located in the right well can be understood as the *difference* between the two lowest energy states.

How should we picture what is going on if we actually place an electron into one well and a second electron into the other well? We said before that we expect the initially empty well to fill with clocks in order to represent the fact that the particle can hop from one side to the other. We even hinted at the answer when we said that the wavefunction 'sloshes' back and forth. To see how this works out, we need to notice that we can express a state localized on one of the protons as the sum of the two lowest-energy wavefunctions. We've illustrated this in Figure 8.2 but what does it mean? If the electron is sat in a particular well at some time, then this implies that it does not actually have a unique energy. Specifically, a measurement of its energy will return a value equal to one of the two possible energies corresponding to the two states of definite energy that build up the wavefunction. The electron is therefore in two energy states at once. We hope that, by this stage in the book, this is not a novel concept.

But here is the interesting thing. Because these two states are not of exactly the same energy, their clocks rotate at slightly different rates (as discussed on page 105). This has the effect that a particle initially described by a wavefunction localized around one proton will, after a long enough time, be described by a wavefunction which is peaked around the other proton. We don't intend to go into details, but suffice to say that this phenomenon is quite analogous to the way that two sound waves of almost the same frequency add together to produce a resultant wave that is at first loud (the two waves are in phase) and then, some time later, quiet (as the two waves are out of phase). This phenomenon is known as 'beats'. As the frequency of the waves gets closer and closer, so the time interval between loud and quiet increases until, when the waves are of exactly the same frequency, they combine to produce a pure tone. This will be completely familiar to any musician who, perhaps unknowingly, exploits this piece of wave physics when they make use of a tuning fork. The story runs in exactly the same way for the second electron sat in the second well. It too tends to migrate from one well to the other in a fashion that exactly mirrors the behaviour of the first electron. Although we might start with one electron in one well and a second electron in the other, after waiting long enough the electrons will swap positions.

We are now going to exploit what we have just learnt. The really interesting physics happens when we start to move the atoms closer together. In our model, moving the atoms together corresponds to reducing the width of the barrier separating the two wells. As the barrier gets thinner, the wavefunctions begin to merge together and the electron is increasingly likely to be found in the region between the two protons. Figure 8.3 illustrates what the four lowest-energy wavefunctions look like when the barrier is thin. It is interesting that the lowest-energy wavefunction is starting to look like the lowest-energy sine wave we would get if we had a single electron in a single, wide well, i.e. the two peaks merge together to produce a single peak (with a dimple in it). Meanwhile, the second-lowest-energy wavefunction looks rather like the sine wave corresponding to the next-to-lowest energy for a single, wide well. This is what we should expect, because, as the barrier between the wells gets thinner, its effect diminishes and, eventually, when it has no thickness at all, it has no effect at all and so our electron should behave exactly as if it is in a single well.

Having looked at what happens at the two extremes – the wells widely separated and the wells close together – we can complete the picture by considering how the allowed electron energies vary as we decrease the distance between the wells. We've sketched the results for the lowest four energy levels in Figure 8.4. Each of the four lines represents one of the four lowest energy levels, and we've sketched the corresponding wavefunctions next to them. The right-hand edge of the picture shows the wavefunctions when the wells are widely separated (see also Figure 8.1). As we expect, the difference between the energy levels of the electrons in each well are virtually indistinguishable. As the wells get closer together, however, the energy levels begin to separate (compare the wavefunctions on the left with those in Figure 8.3). Interestingly, the energy level corresponding to the anti-symmetric wavefunction increases, whilst that corresponding to the symmetric wavefunction decreases.

This has a profound consequence for a real system of two protons and two electrons – that is, two hydrogen atoms. Remember that in reality two electrons can actually fit into the same energy level because they can have opposite spins. This means that they can both fit into the lowest (symmetric) energy level and, crucially, this level decreases in energy as the atoms get closer together. This means that it is energetically favourable for two distant atoms to move closer together. And this is what actually happens in Nature:[4] the symmetric wavefunction describes a system in which the electrons are shared more evenly between the two protons than one might anticipate from the 'far apart' wavefunction, and because this 'sharing' configuration is of lower energy, the atoms are drawn towards each other. This attraction is eventually halted because the two protons are positively charged and as such they repel each other

4. Providing the protons are not moving too rapidly relative to each other.

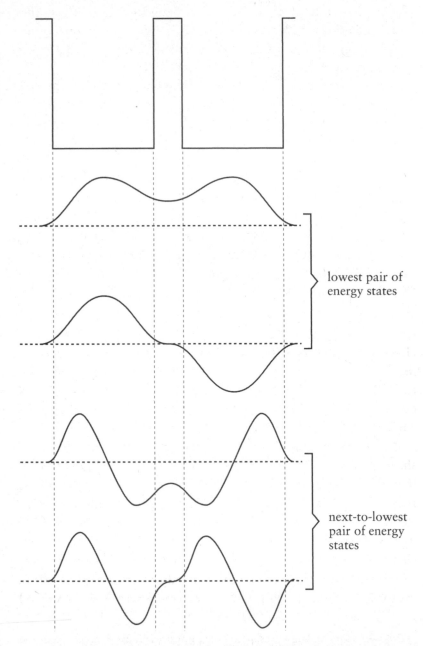

Figure 8.3: Like Figure 8.1 except that the wells are closer together. The 'leakage' into the region in between the wells increases. Unlike Figure 8.1, we also show the wavefunctions corresponding to the pair of next-to-lowest energies.

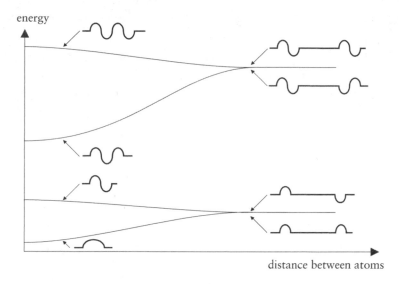

energy

distance between atoms

Figure 8.4: The variation of the allowed electron energies as we change the distance between the wells.

(there is also repulsion due to the fact that the electrons have equal charges), but this repulsion only beats the inter-atomic attraction at distances smaller than around 0.1 nanometres (at room temperature). The result is that a pair of hydrogen atoms at rest will eventually nestle together. This pair of nestled hydrogen atoms has a name: it is a hydrogen molecule.

This preference for two atoms to stick together as a result of sharing their electrons between them is known as a covalent bond. If you look back at the top wavefunction in Figure 8.3, then this is roughly what the covalent bond in a hydrogen molecule looks like. Remember that the height of the wave corresponds to the probability that an electron will be found at that point.[5] There is a peak above each well, i.e. around each proton, which informs us that each electron is still most likely to be in the vicinity of one or other of the protons. But there is also a significant chance that the electrons will

5. This is true for standing waves, where the clock size and the projection onto the 12 o'clock direction are proportional to each other.

spend time between the protons. Chemists speak of the atoms 'sharing' electrons in a covalent bond, and this is what we are seeing, even in our toy model with two square wells. Beyond the hydrogen molecule, this tendency for atoms to share electrons is what we invoked when we were discussing chemical reactions on pages 123–4.

That is a very satisfying conclusion to reach. We have learnt that, for hydrogen atoms that are far apart, the tiny difference between the two lowest-lying energy states was only of academic interest, although it did lead us to conclude that every electron in the Universe knows about every other, which is certainly fascinating. On the other hand, the two states get increasingly separated as the protons get closer together, and the lower of the two eventually becomes the state that describes the hydrogen molecule, and that is very far from being of mere academic interest, because covalent bonding is the reason that you are not a bunch of atoms sloshing around in a featureless blob.

Now we can keep pulling on this intellectual thread and start to think about what happens when we bring more than two atoms together. Three is bigger than two, so let's start there and consider a triple-well potential, as illustrated in Figure 8.5. As ever, we are to imagine that each well is at the site of an atom. There should be three lowest energy states, but looking at the figure you might be tempted to think that there are now four energy states for every state of the single well. The four states we have in mind are illustrated in the figure and they correspond to wavefunctions that are variously symmetric or anti-symmetric about the centre of the two potential barriers.[6] This counting must be wrong, because if it were correct then one could put four identical fermions into these four states and the Pauli principle would be violated. To get the Pauli principle to work out we need just three energy states and this, of course, is

6. You might think there are four more wavefunctions, corresponding to the ones we have drawn turned upside down, but, as we have said, these are equivalent to the ones drawn.

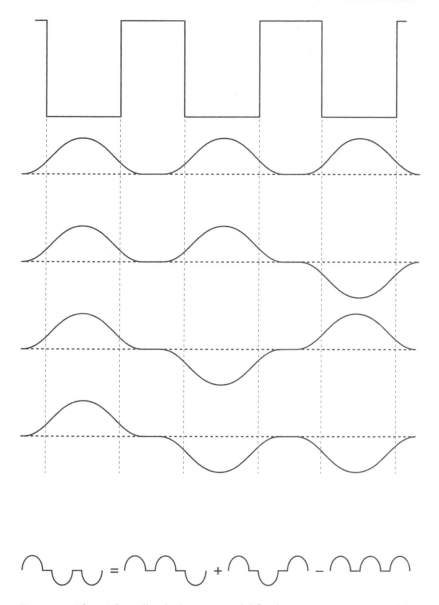

Figure 8.5. The triple well, which is our model for three atoms in a row, and the possible lowest-energy wavefunctions. At the bottom we illustrate how the bottom of the four waves can be obtained from the other three.

what happens. To see this, we need merely spot that we can always write any one of the four wavefunctions sketched in the figure as a combination of the other three. At the bottom of the figure, we have illustrated how that works out in one particular case; we have shown how the last wavefunction can be obtained by a combination of adding and subtracting the other three.

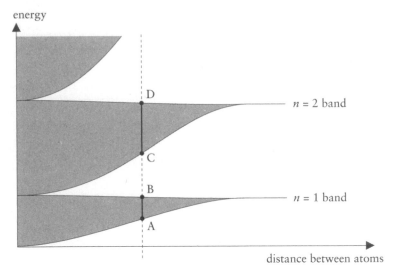

Figure 8.6. The energy bands in a chunk of solid matter and how they vary with the distance between the atoms.

Having identified the three lowest energy states for a particle sitting in the triple-well potential, we can ask what Figure 8.4 looks like in this case, and it should come as no surprise at all to find that it looks rather similar, except that what was a pair of allowed energy states becomes a triplet of allowed states.

Enough of three atoms – we shall now swiftly move our attention to a chain of many. This is going to be particularly interesting because it contains the key ideas that will allow us to explain a lot about what is happening inside solid matter. If there are N wells (to model a chain of N atoms) then for each energy in the single well there will now be N energies. If N is something like 10^{23}, which is typical of the number of atoms in a small chunk of solid material, that is an awful lot of splitting. The result is that Figure 8.4 now looks something like Figure 8.6. The vertical dotted line illustrates that, for atoms that are separated by the corresponding distance, the electrons can only have certain allowed energies. That should be no big surprise (if it is, then you'd better start reading the book again from the beginning), but what is interesting is that the allowed energies come in 'bands'. The energies from A to B are allowed, but no other energies are allowed until we get to C, whence energies from C to D are allowed, and so on. The fact that there are many atoms in the chain means that there are very many allowed energies crammed into each band. So many in fact, that for a typical solid we can just as well suppose that the allowed energies form a smooth continuum in each band. This feature of our toy model is preserved in real solid matter – the electrons there really do have energies that come grouped in bands like this, and that has important implications for what kind of solid we are talking about. In particular, these bands explain why some materials (metals) conduct electricity whilst others (insulators) do not.

How so? Let's begin by considering a chain of atoms (as ever modelled by a chain of potential wells), but now suppose that each atom has several electrons bound to it. This, of course, is the norm – only hydrogen has just the one electron bound to a single proton – and so we are moving from a discussion of a chain of hydrogen atoms to the more interesting case of a chain of heavier atoms. We should also remember that electrons can come in two types; spin up and spin down, and the Pauli principle informs us that we can drop no more than two electrons into each allowed energy level. It follows that for a chain of atoms each containing just one electron per atom (i.e. hydrogen) the $n = 1$ energy band is half-filled. This is illustrated in Figure 8.7, where we have sketched the energy levels for a chain of 5 atoms. This means that each band contains 5 distinct allowed energies. These 5 energy states can accommodate a maximum of 10 electrons, but we only have 5 to worry about so, in the lowest energy configuration, the chain of atoms contains the

5 electrons occupying the bottom half of the $n = 1$ energy band. If we had 100 atoms in the chain then the $n = 1$ band could contain 200 electrons, but for hydrogen, we only have 100 electrons to deal with and so once again the $n = 1$ band is half filled when the chain of atoms is in its lowest energy configuration. Figure 8.7 also shows what happens in the case that there are 2 electrons for every atom (helium) or 3 electrons per atom (lithium). In the case of helium, the lowest-energy configuration corresponds to a filled $n = 1$ band, whilst for lithium the $n = 1$ band is filled and the $n = 2$ band is half filled. It should be pretty clear that this pattern of filled or half-filled continues such that atoms with an even number of electrons always lead to filled bands whilst atoms with an odd number of electrons always lead to half-filled bands. Whether a band is full or not is, as we shall very soon discover, the reason why some materials are conductors whilst others are insulators.

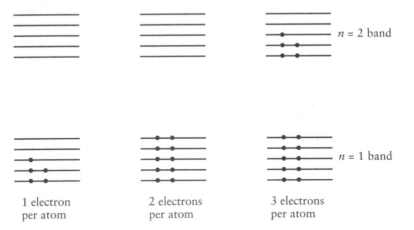

Figure 8.7. The way electrons occupy the lowest available energy states in a chain of five atoms when each atom contains one, two or three electrons. The black dots denote the electrons.

Let's now imagine connecting the ends of our atomic chain to the terminals of a battery. We know from experience that if the atoms form a metal then an electric current will flow. But what does that actually mean, and how does it emerge from our story so far?

The precise action of the battery on the atoms within the wire is, fortunately, something we don't really need to understand. All we need to know is that connecting up the battery provides a source of energy that is able to kick an electron a little, and that kick is always in the same direction. A good question to ask is exactly how a battery does that. To say 'it is because it induces an electric field within the wire and electric fields push electrons' is not entirely satisfying, but it will have to satisfy us as far as this book is concerned. Ultimately, we could appeal to the laws of quantum electrodynamics and try to work the whole thing out in terms of electrons interacting with photons. But we would add absolutely nothing to the current discussion by doing this, so in the interests of brevity, we won't.

Imagine an electron sitting in one of those states of definite energy. We will start by assuming that the action of the battery can only provide very tiny kicks to the electron. If the electron is sat in a low energy state, with many other electrons above it on the energy ladder (we have Figure 8.7 in mind when using this language), it will be unable to receive the energy kick from the battery. It is blocked, because the energy states above it are already filled. For example, the battery might be capable of kicking the electron up to an energy state a few rungs higher, but if all the accessible rungs are already occupied then our target electron must pass up on the opportunity to absorb the energy because there is simply nowhere for it to go. Remember, the Exclusion Principle prevents it from joining the other electrons if the available places are taken. The electron will be forced to behave as if there is no battery connected at all. The situation is different for those electrons with the highest energies. They are lying close to the top of the heap and can potentially absorb a tiny kick from the battery and move into a higher energy state – but only if they are not sitting at the very top of an already full band. Referring back to Figure 8.7, we see that the highest-energy electrons will be able to absorb energy from the battery if the atoms in the chain contain an odd number of electrons. If they contain an even number, then the topmost electrons still cannot go anywhere because there is a big gap in their energy ladder, and they will only overcome this if they are given a large enough kick.

This implies that if the atoms in a particular solid contain an even number of electrons, those electrons may well behave as if the battery had never been connected. A current simply can't flow because there is no way for its electrons to absorb energy. This is a description of an insulator. The only way out of this conclusion is if the gap between the top of the highest filled band and the bottom of the next empty band is sufficiently small – we shall have more to say about that very soon. Conversely, if the atoms contain an odd number of electrons then the topmost electrons are always free to absorb a kick from the battery. As a result they hop up into a higher energy level and, because the kick is always in the same direction, the net effect is to induce a flow of these mobile electrons, which we recognize as an electrical current. Very simplistically, therefore, we might conclude that, if a solid is made up from atoms containing odd numbers of electrons, then they are destined to be conductors of electricity.

Happily, the real world is not that simple. Diamond, a crystalline solid made up entirely of carbon atoms which have six electrons, is an insulator. Graphite, on the other hand, which is also pure carbon, is a conductor. In fact, the odd/even electron rule hardly ever works out in practice, but that is because our 'wells in a line' model of a solid is far too rudimentary. What is absolutely true, though, is that good conductors of electricity are characterized by the fact that the highest-energy electrons have the headroom to leap into higher energy states, whilst insulators are insulators because their topmost electrons are blocked from accessing the higher energy states by a gap in their ladder of allowed energies.

There is a further twist to this tale, and it is a twist that matters when we come to explaining how the current flows in a semiconductor in the next chapter. Let us imagine an electron, free to roam around an unfilled band of a perfect crystal. We say a crystal because we mean to imply that the chemical bonds (possibly covalent) have acted so as to arrange the atoms in a regular pattern. Our one-dimensional model of a solid corresponds to a crystal if all of the wells are equidistant and of the same size. Connect a battery, and an electron will merrily hop up from one level to the next as the applied electric field gently nudges it. As a result, the electric current will steadily increase as the electrons absorb more energy and move faster and faster. To anyone who knows anything about electricity, this should sound rather odd, because there is no sign of 'Ohm's Law', which states that the current (I) should be fixed by the size of the applied voltage (V) according to $V = I \times R$, where R represents the resistance in the wire. Ohm's Law emerges because as the electrons hop their way up the energy ladder they can also lose energy and drop all the way down again – this will only happen if the atomic lattice is not perfect, either because there are impurities within the lattice (i.e. rogue atoms that are different from the majority) or if the atoms are jiggling around significantly, which is what is guaranteed to happen at any non-zero temperature. As a result, the electrons spend most of their time playing a microscopic game of snakes and ladders as they climb up the energy ladder only to fall down again as a result of their interactions with the less than perfect atomic lattice. The average effect is to produce a 'typical' electron energy and that leads to a fixed current. This typical electron energy determines how fast the electrons flow down the wire and that is what we mean by a current of electricity. The resistance of the wire is to be seen as a measure of how imperfect the atomic lattice is through which the electrons are moving.

But that is not the twist. Even without Ohm's Law, the current doesn't just keep increasing. When electrons reach the top of a band, they behave very oddly indeed, and the net effect of this behaviour is to decrease the current and eventually reverse it. This is very odd: even though the electric field is kicking the electrons in one direction, they end up travelling in the opposite direction when they near the top of a band. The explanation of this weird effect is beyond the scope of this book, so we shall just say that the role of the positively charged atomic cores is the key, and they act to push the electrons so that they reverse direction.

Now, as advertised, we will explore what happens when a would-be insulator behaves like a conductor because the gap between the last filled band and the next, empty, band is 'sufficiently small'. At this stage it is worth introducing some jargon. The last (i.e. highest-energy) band of energies that is completely filled with electrons is referred to as the 'valence band', and the next band up (either empty or half-filled in our analysis) is referred to as the 'conduction band'. If the valence and conduction bands actually overlap (and that is a real possibility), then there is no gap at all and a would-be insulator instead behaves as a conductor. What if there is a gap but the gap is 'sufficiently small'? We have indicated that the electrons can receive energy from a battery, so we might suppose that, if the battery is powerful, then it could deliver a mighty enough kick to project an electron sitting near to the top of the valence band up into the conduction band. That is possible, but this is not where our interest lies because typical batteries can't generate a big enough kick. To put some numbers on it, the electric field within a solid is typically of the order of a few volts per metre, and we would need fields of a few volts per nanometre (i.e. a billion times stronger) in order to provide a kick capable of making an electron jump the electron volt[7] or so in energy needed to leap from the valence band to the conduction band in a typical insulator. Much more interesting is the kick that an electron can receive from the atoms that make up the solid. They are not rigidly sitting in the same place, but rather they are jiggling around a little bit – the hotter the solid the more they jiggle and a jiggling atom can deliver far more energy to an electron than a practical battery; enough to make it leap a few electron volts in energy. At room temperature, it is actually very rare to hit an electron that

---

7. The electron volt is a very convenient unit of energy for discussing electrons in atoms and is widely used in nuclear and particle physics. It is the energy an electron would acquire if it were accelerated through a potential difference of 1 volt. That definition is not important, all that matters is that it is a way of quantifying energy. To get a feel for the size, the energy required to completely liberate an electron from the ground state of a hydrogen atom is 13.6 electron volts.

hard, because at 20°C the typical thermal energies are around $\frac{1}{40}$ of an electron volt. But this is only an average, and there are a very large number of atoms in a solid, so it does occasionally happen. When it does, electrons can leap from their valence band prison into the conduction band, where they may then absorb the tiny kicks from a battery and in so doing initiate a flow of electricity.

Materials in which, at room temperature, a sufficient number of electrons can be lifted up from the valence to conduction band in this way have their own special name: they are called semiconductors. At room temperature they can carry a current of electricity, but as they are cooled down, and their atoms jiggle less, so their ability to conduct electricity diminishes and they turn back into insulators. Silicon and germanium are the two classic examples of semiconductor materials and, because of their dual nature, they can be used to great advantage. Indeed, it is no exaggeration to say that the technological application of semiconductor materials revolutionized the world.

9. The Modern World

In 1947, the world's first transistor was built. Today, every year the world manufactures over 10,000,000,000,000,000,000,000, which is one hundred times more than the sum total of all the grains of rice consumed every year by the world's seven billion residents. The world's first transistor computer was built in Manchester in 1953, and had ninety-two of them. Today, you can buy over a hundred thousand transistors for the cost of a single grain of rice and there are around a billion of them in your mobile phone. In this chapter, we are going to describe how a transistor works, surely the most important application of quantum theory.

As we saw in the previous chapter, a conductor is a conductor because some of the electrons are sitting in the conduction band. As a result, they are quite mobile and can 'flow down' the wire when a battery is connected. The analogy with flowing water is a good one; the battery is causing current to flow. We can even use the 'potential' concept to capture this idea, because the battery creates a potential within which the conduction electrons move, and the potential is in a sense, 'downhill'. So an electron in the conduction band of a material 'rolls' down the potential created by the battery, gaining energy as it goes. This is another way to think about the tiny kicks we talked about in the last chapter – instead of a battery inducing tiny kicks that accelerate the electron along the wire, we are invoking a classical analogy akin to water flowing down a hill. This is a good way to think about the conduction of electricity by electrons, and it is the way we will be thinking throughout the rest of this chapter.

In a semiconductor material like silicon, something very interesting happens because the current is not only carried by electrons in the conduction band. The electrons in the valence band contribute to

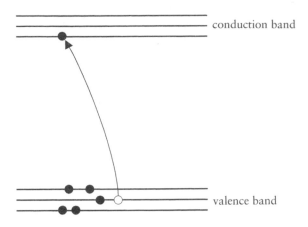

Figure 9.1. An electron-hole pair in a semiconductor.

the current too. To see that, take a look at Figure 9.1. The arrow shows an electron, originally sitting inert in the valence band, absorbing some energy and being lifted up into the conduction band. Certainly the elevated electron is now much more mobile, but something else is mobile too – there is now a hole left in the valence band, and that hole provides some wriggle room for the otherwise inert valence band electrons. As we have seen, connecting a battery to this semiconductor will cause the conduction band electron to hop up in energy, thereby inducing an electric current. What happens to that hole? The electric field created by the battery can cause an electron from some lower energy state in the valence band to hop into the vacant hole. The hole is filled in, but now there is a hole 'deeper' down in the valence band. When electrons in the valence band hop into the vacant hole, the hole moves around.

Rather than bother keeping track of the motion of all the electrons in the almost-full valence band, we can instead decide to keep track of where the hole is and forget about the electrons. That book-keeping convenience is the norm for those working on the physics of semiconductors, and it will make our life simpler to think in that way too.

An applied electric field induces the conduction band electrons to flow, creating a current, and we should like to know what it does to the holes in the valence band. We know that the valence band electrons are not free to move, because they are almost completely trapped by the Pauli principle but they will shuffle along under the influence of the electric field and the hole moves along with them. This might sound counterintuitive, and if you are having trouble with the idea that if electrons in the valence band shuffle to the left then the hole also shuffles to the left, perhaps the following analogy will help. Imagine a line of people all standing in a queue 1 metre apart, except that somewhere in the middle of the line a single person is missing. The people are analogous to electrons and the missing person is the hole. Now imagine that all the people stride 1 metre forwards so that they end up where the person in front of them was standing. Obviously the gap in the line jumps 1 metre forwards too, and so it is with the holes. One could also imagine water flowing down a pipe – a small bubble in the water will move in the same direction as the water, and this 'missing water' is analogous to a hole in the valence band.

But, as if that wasn't enough to be going on with, there is an important added complication; we now need to invoke the piece of physics that we introduced in the 'twist' at the end of the last chapter. If you recall, we said that electrons moving near to the top of a filled band are accelerated by an electric field in the *opposite* direction to electrons moving near to the bottom of a band. This means that the holes, which are near the top of the valence band, move in the opposite direction to the electrons, which are near the bottom of the conduction band.

The bottom line is that we can picture a flow of electrons in one direction and a corresponding flow of holes in the other direction. A hole can be thought of as carrying an electric charge that is exactly opposite to the charge of an electron. To see this, remember that the material through which our electrons and holes flow is, on average, electrically neutral. In any ordinary region there is no net charge, because the charge due to the electrons cancels the positive charge carried by the atomic cores. But if we make an electron–hole pair by exciting an electron out of the valence band and into the conduction band (as we have been discussing), then there is a free electron roaming around, which constitutes an excess of negative charge relative to the average conditions in that region of the material. Likewise, the hole is a place where there is no electron and so it corresponds to a region where there is a net excess of positive charge. The electric current is defined to be the rate at which positive charges flow,[1] and so electrons contribute negatively to the current and the holes contribute positively, if they are flowing in the same direction. If, as is the case in our semiconductor, the electrons and holes flow in opposite directions, then the two add together to produce a larger net flow of charge and hence a larger current.

Whilst all this is a little intricate, the net effect is very straightforward: we are to imagine a current of electricity through a semiconductor material as being representative of the flow of charge, and this flow can be made up of conduction band electrons moving in one direction and valence band holes moving in the opposite direction. This is to be contrasted with the flow of current in a conductor – in that case, the current is dominated by the flow of a large number of electrons in the conduction band, and the extra current coming from electron–hole pair production is negligible.

To understand the utility of semiconductor materials is to appreciate that the current flowing in a semiconductor is not like an uncontrollable flood of electrons down a wire, as it is in a conductor. Instead, it is a much more delicate combination of electron and hole currents and, with a little clever engineering, that delicate combination can be exploited to produce tiny devices that are capable of exquisitely controlling the flow of current through a circuit.

What follows is an inspiring example of applied physics and engineering. The idea is to deliberately contaminate a piece of pure

---

1. This definition is purely a matter of convention and a historical curiosity. We could just as well define the current to flow in the direction that the conduction band electrons move.

silicon or germanium so as to induce some new available energy levels for the electrons. These new levels will allow us to control the flow of electrons and holes through our semiconductor just as we might control the flow of water through a network of pipes using valves. Of course, anyone can control the flow of electricity through a wire – just pull the plug. But that is not what we are talking about – rather we are talking about making tiny switches that allow the current to be controlled dynamically within a circuit. Tiny switches are the building blocks of logic gates, and logic gates are the building blocks of microprocessors. So how does that all work out?

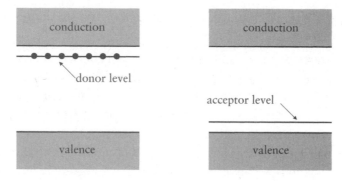

Figure 9.2. The new energy levels induced in a n-type semiconductor (on the left) and a p-type semiconductor (on the right).

The left-hand part of Figure 9.2 illustrates what happens if a piece of silicon is contaminated with phosphorous. The degree of contamination can be controlled with precision and this is very important. Suppose that every now and then within a crystal of pure silicon an atom is removed and replaced with a phosphorous atom. The phosphorous atom snuggles neatly into the spot vacated by the silicon atom, the only difference being that phosphorous has one *more* electron than silicon. That extra electron is very weakly bound to its host atom, but it is not entirely free and so it occupies an energy level lying just below the conduction band. At low temperatures the conduction band is empty, and the extra electrons donated by the phosphorous atoms reside in the donor level marked in the figure.

At room temperature, electron–hole pair creation in the silicon is very rare, and only about one electron in every trillion gets enough energy from the thermal vibrations of the lattice to jump out of the valence band and into the conduction band. In contrast, because the donor electron in phosphorous is so weakly bound to its host, it is very likely that it will make the small hop from the donor level into the conduction band. So at room temperature, for levels of doping greater than one phosphorous atom for every trillion silicon atoms, the conduction band will be dominated by the presence of the electrons donated by the phosphorous atoms. This means it is possible to control very precisely the number of mobile electrons that are available to conduct electricity, simply by varying the degree of phosphorous contamination. Because it is electrons roaming in the conduction band that are free to carry the current, we say that this type of contaminated silicon is 'n-type' ('n' for 'negatively charged').

The right-hand part of Figure 9.2 shows what happens if instead we contaminate the silicon with atoms of aluminium. Again, the aluminium atoms are sprinkled sparingly around among the silicon atoms, and again they snuggle nicely into the spaces where silicon atoms would otherwise be. The difference comes because aluminium has one *fewer* electron than silicon. This introduces holes into the otherwise pure crystal, just as phosphorous added electrons. These holes are located in the vicinity of the aluminium atoms, and they can be filled in by electrons hopping out of the valence band of neighbouring silicon atoms. The 'hole-filled' acceptor level is illustrated in the figure, and it sits just above the valence band because it is easy for a valence electron in the silicon to hop into the hole made by the aluminium atom. In this case, we can naturally regard the electric current as being propagated by the holes, and for that reason this kind of contaminated silicon is known as 'p-type' ('p' for 'positively charged'). As before, at room temperature, the level of aluminium contamination does not need to be much more than one part per trillion before the current due to the motion of the holes from the aluminium is dominant.

So far we have simply said that it is possible to make a lump of

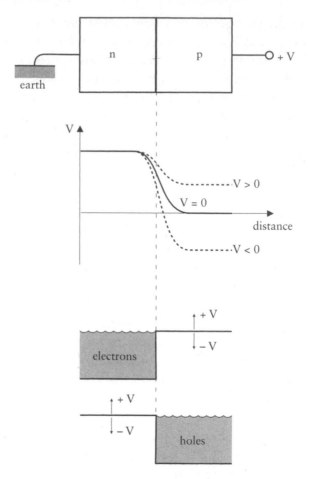

Figure 9.3. A junction formed by joining together a piece of n-type and a piece of p-type silicon.

silicon which is able to transmit a current, either by allowing electrons donated by phosphorous atoms to sail along in the conduction band or by allowing holes donated by aluminium atoms to sail along in the valence band. What is the big deal?

Figure 9.3 illustrates that we are on to something because it shows what happens if we join together two pieces of silicon; one n-type and the other p-type. Initially, the n-type region is awash with electrons from the phosphorous and the p-type region is awash with holes from the aluminium. As a result, electrons from the n-type region drift over into the p-type region, and holes from the p-type region drift over into the n-type region. There is nothing mysterious about this; the electrons and holes simply meander across the junction between the two materials just as a drop of ink spreads out in a bath of water. But as the electrons and holes drift in opposite directions, they leave behind regions of net positive charge (in the n-type region) and net negative charge (in the p-type region). This build up of charge opposes further migration by the 'like sign charges repel' rule, until eventually there is a balance, and no further net migration occurs.

The second of the three pictures in Figure 9.3 illustrates how we might think of this using the language of potentials. What is shown is how the electric potential varies across the junction. Deep in the n-type region, the effect of the junction is unimportant, and since the junction has settled into a state of equilibrium, no current flows. That means the potential is constant inside this region. Before moving on we should once again be clear what the potential is doing for us: it is simply telling us what forces act on the electrons and holes. If the potential is flat, then, just as a ball sitting on flat ground will not roll, an electron will not move.

If the potential dips down then we might suppose that an electron placed in the vicinity of the falling potential will 'roll downhill'. Inconveniently, convention has it the other way and a downhill potential means 'uphill' for an electron, i.e. electrons will flow uphill. In other words, a falling potential acts as a barrier to an electron, and this is what we've drawn in the figure. There is a force pushing the electron away from the p-type region as a result of the build up of negative charge that has occurred by earlier electron migration. This force is what prevents any further net migration of electrons from the n-type to the p-type silicon. Using downhill potentials to represent an uphill journey for an electron is actually not as silly as it seems, because things now make sense from the point of view of the holes, i.e. holes naturally flow downhill. So now we can also see that the way we drew the potential (i.e. going from the high ground on the left to low ground on the right) also correctly accounts for the fact that holes are prevented from escaping from the p-type region by the step in the potential.

The third picture in the figure illustrates the flowing water analogy. The electrons on the left are ready and willing to flow down the wire but they are prevented from doing so by a barrier. Likewise the holes in the p-type region are stranded on the wrong side of the barrier; the water barrier and the step in the potential are just two different ways of speaking about the same thing. This is how things are if we simply stick together an n-type piece of silicon and a p-type piece. Actually, the act of sticking them together takes more care than we are suggesting – the two cannot simply be glued together, because then the junction will not allow the electrons and holes to flow freely from one region to the other.

Interesting things start to happen if we now connect this 'pn junction' up to a battery, which allows us to raise or lower the potential barrier between the n-type and p-type regions. If we lower the potential of the p-type region then we steepen the step and make it even harder for the electrons and holes to flow across the junction. But raising the potential of the p-type region (or lowering the potential of the n-type region) is just like lowering the dam that was holding back the water. Immediately, electrons will flood from n-type to p-type and holes will flood in the opposite direction. In this way a pn-junction can be used as a diode – it can allow a current to flow, but only in one direction. Diodes are, however, not where our ultimate interest resides.

Figure 9.4 is a sketch of the device that changed the world – the transistor. It shows what happens if we make a sandwich, with a layer of p-type silicon in between two layers of n-type silicon. Our explanation of a diode will serve us well here, because the ideas are basically the same. Electrons drift from the n-type regions into the p-type region and holes drift the other way until this diffusion is eventually halted by the potential steps at the junctions between the layers. In isolation, it is as if there are two reservoirs of electrons held apart by a barrier, and a single reservoir of holes that sits brim-full in between.

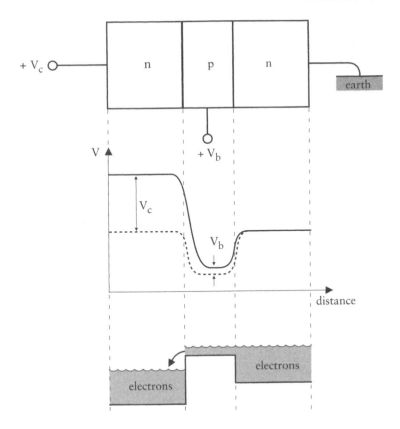

Figure 9.4: A transistor.

The interesting action occurs when we apply voltages to the n-type region on one side and the p-type region in the middle. Applying positive voltages causes the plateau on the left to rise (by an amount V_c) and likewise the plateau in the p-type region (by an amount V_b). We've indicated this by the solid line in the middle diagram in the figure. This way of arranging the potentials has a dramatic effect, because it creates a waterfall of electrons as they flood over the lowered central barrier and into the n-type region on the left (remember, electrons flow 'uphill'). Providing that V_c is larger than V_b, the flow of electrons is one-way and the electrons on the left remain unable to flow across the p-type region. This all might sound rather innocuous, but we have just described an electronic valve. By applying a voltage to the p-type region we are able to turn on and off the electron current.

Now comes the finale – we are ready to recognize the full potential of the humble transistor. In Figure 9.5 we illustrate the action of a transistor by once again drawing parallels with flowing water. The 'valve closed' situation is entirely analogous to what happens if no voltage is applied to the p-type region. Applying a voltage corresponds to opening up the valve. Below the two pipes, we have also drawn the symbol that is often used to represent a transistor and, with a little imagination, it even looks a little like a valve.

What can we do with valves and pipes? The answer is that we can

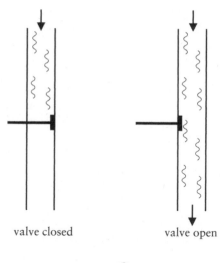

valve closed   valve open

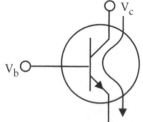

Figure 9.5. The 'water in a pipe' analogy with a transistor.

build a computer and if those pipes and valves can be made small enough then we can make a serious computer. Figure 9.6 illustrates conceptually how we can use a pipe with two valves to construct something called a 'logic gate'. The pipe on the left has both valves open and as a result water flows out of the bottom. The pipe in the middle and the pipe on the right both have one valve closed and obviously no water can then flow out of the bottom. We have not bothered to show the fourth possibility, when both valves are closed. If we were to represent the flow of water out of the bottom of our pipes by the digit '1' and the absence of flow by the digit '0', and if we assign the digit '1' to an open valve and the digit '0' to a closed valve, then we can summarize the action of the four pipes (three drawn and one not) by the equations '1 AND 1 = 1', '1 AND 0 = 0', '0 AND 1 = 0' and '0 AND 0 = 0'. The word 'AND' is here a logical operation and it is being used in a technical way – the system of pipe and valves we just described is called an 'AND gate'. The gate takes two inputs (the state of the two valves) and returns a single output (whether water flows or not) and the only way to get a '1' out is to feed a '1' and a '1' in. We hope it is clear how we can use a pair

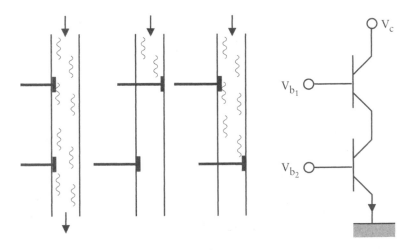

Figure 9.6. An 'AND' gate built using a water pipe and two valves (*left*) or a pair of transistors (*right*). The latter is much better suited to building computers.

of transistors connected in series to built an AND gate – the circuit diagram is illustrated in the figure. We see that only if both transistors are turned on (i.e. by applying positive voltages to the p-type regions, V_{b_1} and V_{b_2}) is it possible for a current to flow, which is just what is needed to implement an AND gate.

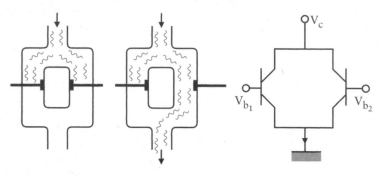

Figure 9.7. An 'OR' gate built using water pipes and two valves (*left*) or a pair of transistors (*right*).

Figure 9.7 illustrates a different logic gate. This time water will flow out of the bottom if either valve is open and only if both are closed will it not flow. This is called an 'OR' gate and, using the same notation as before, '1 OR 1 = 1', '1 OR 0 = 1', '0 OR 1 = 1' and '0 OR 0 = 0'. The corresponding transistor circuit is also illustrated in the figure and now a current will flow in all cases except when both transistors are switched off.

Logic gates like these are the secret behind the power of digital electronic devices. Starting from these modest building blocks one can assemble combinations of logic gates in order to implement arbitrarily sophisticated algorithms. We can imagine specifying a list of inputs into some logical circuits (a series of '0's and '1's), sending these inputs through some sophisticated configuration of transistors that spits out a list of outputs (again a series of '0's and '1's). In that way we can build circuits to perform complicated mathematical calculations, or to make decisions based on which keys are pressed on a keyboard, and feed that information to a unit which then displays the corresponding characters on a screen, or to trigger an alarm if an intruder breaks into a house, or to send a stream of text characters down a fibre optic cable (encoded as a series of binary digits) to the other side of the world, or . . . in fact, anything you can think of, because virtually every electrical device you possess is crammed full of transistors.

The potential is limitless, and we have already exploited the transistor to change the world enormously. It is probably not overstating things to say that the transistor is the most important invention of the last 100 years – the modern world is built on and shaped by semiconductor technologies. On a practical level, these technologies have saved millions of lives – we might point in particular to the applications of computing devices in hospitals, the benefits of rapid, reliable and global communication systems and the uses of computers in scientific research and in controlling complex industrial processes.

William B. Shockley, John Bardeen and Walter H. Brattain were awarded the Nobel Prize in Physics in 1956 'for their researches on semiconductors and their discovery of the transistor effect'. There has probably never been a Nobel Prize awarded for work that directly touches so many people's lives.

10.   *Interaction*

In the opening chapters we set up the framework to explain how tiny particles move around. They hop around, exploring the vastness of space without any prejudice, metaphorically carrying their tiny clocks with them as they go. When we add together the multitude of clocks corresponding to the different ways that a particle can arrive at some particular point in space, we obtain one definitive clock whose size informs us of the chance of finding the particle 'there'. From this wild, anarchic display of quantum leaping emerges the more familiar properties of everyday objects. In a sense, every electron, every proton and every neutron in your body is constantly exploring the Universe at large, and only when the sum total of all those explorations is computed do we arrive at a world in which the atoms in your body, fortunately, tend to stay in a reasonably stable arrangement – at least for a century or so. What we have not yet addressed in any detail is the nature of the interactions between particles. We have managed to make a lot of progress without being specific about how particles actually talk to each other, in particular by exploiting the concept of a potential. But what is a potential? If the world is made up solely of particles, then surely we should be able to replace the vague notion that particles move 'in the potential' created by other particles, and speak instead about how the particles move and interact with each other.

The modern approach to fundamental physics, known as quantum field theory, does just this by supplementing the rules for how particles hop around with a new set of rules that explain how those particles interact with each other. These rules turn out to be no more complicated than the rules we've met so far, and it is one of the wonders of modern science that, despite the intricate complexity of the natural world, there are not many of them. 'The eternal mystery of the world is its comprehensibility,' Albert Einstein wrote, and 'the fact that it is comprehensible is a miracle.'

Let's start by articulating the rules of the first quantum field theory to be discovered – quantum electrodynamics, or QED. The origins of the theory can be traced all the way back to the 1920s, when Dirac in particular had an initial burst of success in quantizing Maxwell's electromagnetic field. We've already met the quantum of the electromagnetic field many times in this book – it is the photon – but there were many problems associated with the new theory that were apparent but remained unsolved throughout the 1920s and 1930s. How exactly does an electron emit a photon when it moves between the energy levels in an atom, for example? And, for that matter, what happens to a photon when it is absorbed by an electron, allowing the electron to jump to a higher energy level in the first place? Photons can obviously be created and destroyed in atomic processes, and the means by which this happens is not addressed in the 'old-fashioned' quantum theory that we have met so far in this book.

In the history of science, there are a handful of legendary gatherings of scientists – meetings that certainly appear to have changed the course of science. They probably didn't, in the sense that the participants had usually been working on problems for years, but the Shelter Island Conference of June 1947, held at the tip of Long Island, New York, has a better claim than most for catalysing something special. The participant list alone is worth reciting, because it is short and yet a role-call of the greats of twentieth-century American physics. In alphabetical order: Hans Bethe, David Bohm, Gregory Breit, Karl Darrow, Herman Feshbach, Richard Feynman, Hendrik Kramers, Willis Lamb, Duncan MacInnes, Robert Marshak, John von Neumann, Arnold Nordsieck, J. Robert Oppenheimer, Abraham Pais, Linus Pauling, Isidor Rabi, Bruno Rossi, Julian Schwinger, Robert Serber, Edward Teller, George Uhlenbeck, John Hasbrouck van Vleck, Victor Weisskopf and John Archibald Wheeler. The reader has met several of these names in this book already, and any student of physics probably has heard of most of them. The American writer Dave

Barry once wrote: 'If you had to identify, in one word, the reason why the human race has not achieved, and never will achieve, its full potential, that word would be meetings.' This is doubtless true, but Shelter Island was an exception. The meeting began with a presentation of what has become known as the Lamb shift. Willis Lamb, using high-precision microwave techniques developed during the Second World War, found that the hydrogen spectrum was not, in fact, perfectly described by old-fashioned quantum theory. There was a minute shift in the observed energy levels that could not be accounted for using the theory we have developed so far in this book. It is a tiny effect, but it was a wonderful challenge to the assembled theorists.

We shall leave Shelter Island there, poised after Lamb's talk, and turn to the theory that emerged in the months and years that followed. In doing so we will uncover the origin of the Lamb shift, but, to whet your appetite, here is a cryptic statement of the answer: the proton and electron are not alone inside the hydrogen atom.

QED is the theory that explains how electrically charged particles, like electrons, interact with each other and with particles of light (photons). It is single-handedly capable of explaining all natural phenomena with the exception of gravity and nuclear phenomena. We'll turn our attention to nuclear phenomena later on, and in doing so explain why the atomic nucleus can hold together even though it is a bunch of positively charged protons and zero charge neutrons which would fly apart in an electro-repulsive instant without some sub-nuclear goings-on. Pretty much everything else – certainly everything you see and feel around you – is explained at the deepest known level by QED. Matter, light, electricity and magnetism – it is all QED.

Let's begin by exploring a system we have already met many times throughout the book: a world containing one single electron. The little circles in the 'clock hopping' figure on page 50 illustrate the various possible locations of the electron at some instant in time. To deduce the likelihood of finding it at some point X at a later time, our quantum rules say that we are to allow the electron to hop to X from every possible starting point. Each hop delivers a clock to X, we add up these clocks and then we are done.

We're going to do something now that might look a little over-complicated at first, but of course there is a very good reason. It's going to involve a few As, Bs and Ts – in other words we're heading off into the land of tweed jackets and chalk dust again; it won't last long.

When a particle goes from a point A at time zero to a point B at time T, we can calculate what the clock at B will look like by winding the clock at A backwards by an amount determined by the distance of B from A and the time interval, T. In shorthand, we can write that the clock at B is given by $C(A,0)P(A,B,T)$ where $C(A,0)$ represents the original clock at A at time zero and $P(A,B,T)$ embodies the clock-winding and shrinking rule associated with the leap from A to B.[1] We shall refer to $P(A,B,T)$ as the 'propagator' from A to B. Once we know the rule of propagation from A to B, then we are all set and can figure out the probability to find the particle at X. For the example in Figure 4.2, we have lots of starting points so we'll have to propagate from every one of them to X, and add all the resulting clocks up. In our seemingly overkill notation, the resultant clock $C(X,T) = C(X1,0)P(X1,X,T) + C(X2,0)P(X2,X,T) + C(X3,0)P(X3,X,T) + \ldots$ where $X1$, $X2$, $X3$, etc. label all the positions of the particle at time zero (i.e. the positions of the little circles in Figure 4.2). Just to be crystal clear, $C(X3,0)P(X3,X,T)$ simply means 'take a clock from point $X3$ and propagate it to point X at time T'. Don't be fooled into thinking there is something tricky going on. All we are doing is writing down in a fancy shorthand something we already knew: 'take the clock at $X3$ and time zero and figure out by how much to turn and shrink it corresponding to the particle making the journey from $X3$ to X at some time T later and then repeat that for all of the other time-zero clocks and finally add all of the clocks together according to the clock-adding rule'. We're

1. The propagator shrinks the clock as well, in order to make sure that the particle will be found with a probability of 1 somewhere in the Universe at time T.

sure you'll agree that this is a bit of a mouthful, and the little bit of notation makes life easier.

We can certainly think of the propagator as the embodiment of the clock-winding and shrinking rule. We can also think of it as a clock. To clarify that bald statement, imagine if we know for certain that an electron is located at point A at time $T = 0$, and that it is described by a clock of size 1 reading 12 o'clock. We can picture the act of propagation using a second clock whose size is the amount that the original clock needs to be shrunk and whose time encodes the amount of winding we need. If a hop from A to B requires shrinking the initial clock by a factor of 5 and winding back by 2 hours, then the propagator $P(A,B,T)$ could be represented by a clock whose size is $\frac{1}{5} = 0.2$ and which reads 10 o'clock (i.e. it is wound 2 hours back from 12 o'clock). The clock at B is simply obtained by 'multiplying' the original clock at A by the propagator clock.

As an aside for those who know about complex numbers, just as each of the $C(X1,0)$, $C(X2,0)$ can be represented by a complex number so can the $P(X1,X,T)$, $P(X2,X,T)$ and they are combined according to the mathematical rules for multiplying two complex numbers together. For those who do not know about complex numbers: it doesn't matter because the description in terms of clocks is equally accurate. All we did was introduce a slightly different way of thinking about the clock-winding rule: we can wind and shrink a clock using another clock.

We are free to design our clock multiplication rule to make this all work: multiply the sizes of the two clocks ($1 \times 0.2 = 0.2$) and combine the times on the two clocks such that we wind the first clock backwards by 12 o'clock minus 10 o'clock = 2 hours. This does sound a little bit like we are over-elaborating, and it is clearly not necessary when we only have one particle to think about. But physicists are lazy, and they wouldn't go to all this trouble unless it saved time in the long run. This little bit of notation proves to be a very useful way of keeping track of all the winding and shrinking when we come to the more interesting case where there are multiple particles in the problem – the hydrogen atom, for example.

Regardless of the details, there are just two key elements in our method of figuring out the chances to find a lone particle somewhere in the Universe. First, we need to specify the array of initial clocks which codify the information about where the particle is likely to be found at time zero. Second, we need to know the propagator $P(A,B,T)$, which is itself a clock encoding the rule for shrinking and turning as a particle leaps from A to B. Once we know what the propagator looks like for any pair of start and end points then we know everything there is to know, and we can confidently figure out the magnificently dull dynamics of a Universe containing a single particle. But we should not be so disparaging, because this simple state of affairs doesn't get much more complicated when we add particle interactions into the game. So let's do that now.

Figure 10.1 illustrates pictorially all of the key ideas we want to discuss. It is our first encounter with Feynman diagrams, the calculational tool of the professional particle physicist. The task we are charged with is to work out the probability of finding a pair of electrons at the points X and Y at some time T. As our starting point we are told where the electrons are at time zero, i.e. we are told what their initial clock clusters look like. This is important because being able to answer this type of question is tantamount to being able to know 'what happens in a Universe containing two electrons'. That may not sound like much progress, but once we have figured this out the world is our oyster, because we will know how the basic building blocks of Nature interact with each other.

To simplify the picture, we've drawn only one dimension in space, and time advances from left to right. This won't affect our conclusions at all. Let's start out by describing the first of the series of pictures in Figure 10.1. The little dots at $T = 0$ correspond to the possible locations of the two electrons at time zero. For the purposes of illustration, we've assumed that the upper electron can be in one of three locations, whilst the lower is in one of two locations (in the real world we must deal with electrons that can be located in an infinity of possible locations, but we'd run out of ink if we had to draw that). The upper electron hops to A at some later time whereupon it does

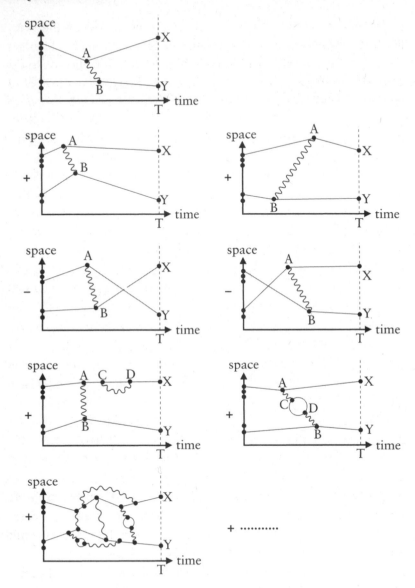

Figure 10.1. Some of the ways that a pair of electrons can scatter off each other. The electrons start out on the left and always end up at the same pair of points, X and Y, at time T. These graphs correspond to some of the different ways that the particles can reach X and Y.

something interesting: it emits a photon (represented by the wavy line). The photon then hops to B where it gets absorbed by the other electron. The upper electron then hops from A to X whilst the lower electron hops from B to Y. That is just one of an infinite number of ways that our original pair of electrons can make their way to points X and Y. We can associate a clock with this entire process – let's call it 'clock 1' or C1 for short. The job of QED is to provide us with the rules of the game that will allow us to deduce this clock.

Before getting into the details, let's sketch how this is going to pan out. The uppermost picture represents one of the myriad ways that the initial pair of electrons can make their way to X and Y. The other pictures represent some of those ways. The crucial idea is that for each possible way that the electrons can get to X and Y we are to identify a quantum clock – C1 is the first in a long list of clocks.[2] When we've got all of the clocks, we are to add them all together and obtain one 'master' clock. The size of that clock (squared) tells us the probability of finding the pair of electrons at X and Y. So once again we are to imagine that the electrons make their way to X and Y not by one particular route, but rather by scattering off each other in every possible way. If we look at the final few pictures in the figure, we can see a variety of more elaborate ways for the electrons to scatter. The electrons not only exchange photons, they can emit and reabsorb a photon themselves, and in the final two figures something very odd is happening. These pictures include the scenario where a photon appears to emit an electron which 'goes in a circle' before ending up where it started out – we shall have more to say about that in a little while. For now, we can simply imagine a series of increasingly complicated diagrams corresponding to cases where the electrons emit and absorb huge numbers of photons before finally ending up at X and Y. We'll need to contemplate the multifarious ways that the electrons can end up at X and Y, but there are two very clear rules: *electrons*

2. We met this idea before, when we tackled the Pauli Exclusion Principle in Chapter 7.

can only hop from place to place and emit or absorb a single photon. That's really all there is to it; electrons can hop or they can branch. Closer inspection should reveal that none of the pictures we have drawn contravenes those two rules because they never involve anything more complicated than a junction involving two electrons and a photon. We must now explain how to go about computing the corresponding clocks, one for each picture in Figure 10.1.

Let's focus on the uppermost picture and explain how to determine what the clock associated with it (clock C1) looks like. Right at the start of the process, there are two electrons sitting there, and they will each have a clock. We should start out by multiplying them together according to the clock multiplication rule to get a new, single clock, which we will denote by the symbol C. Multiplying them makes sense because we should remember that the clocks are actually encoding probabilities, and if we have two independent probabilities then the way to combine them is to multiply them together. For example, the probability that two coins will both come up heads is $\frac{1}{2} \times \frac{1}{2} = \frac{1}{4}$. Likewise, the combined clock, C, tells us the probability to find the two electrons at their initial locations.

The rest is just more clock multiplication. The upper electron hops to A, so there is a clock associated with that; let's call it P(1,A) (i.e. 'particle 1 hops to A'). Meanwhile the lower electron hops to B and we have a clock for that too; call it P(2,B). Likewise there are two more clocks corresponding to the electrons hopping to their final destinations; we shall denote them by P(A,X) and P(B,Y). Finally, we also have a clock associated with the photon, which hops from A to B. Since the photon is not an electron, the rule for photon propagation could be different for the rule for electron propagation so we should use a different symbol for its clock. Let's denote the clock corresponding to the photon hop L(A,B).[3] Now

3. This is a technical point because the clock-winding and -shrinking rule we have used throughout the book to this point does not include the effects of Special Relativity. Including these, as we always must if we are to describe photons, means that the clock-winding rules are different for the electron and photon.

we simply multiply all the clocks together to produce one 'master' clock: $R = C \times P(1,A) \times P(2,B) \times P(A,X) \times P(B,Y) \times L(A,B)$. We are very nearly done now, but there remains some additional clock shrinking to do because the QED rule for what happens when an electron emits or absorbs a photon says that we should introduce a shrinking factor, g. In our diagram, the upper electron emits the photon and the lower one absorbs it – that makes for two factors of g, i.e. g^2. Now we really are done and our final 'clock 1' is obtained by computing $C1 = g^2 \times R$.

The shrinking factor g looks a bit arbitrary, but it has a very important physical interpretation. It is evidently related to the probability that an electron will emit a photon, and this encodes the strength of the electromagnetic force. Somewhere in our calculation we had to introduce a connection with the real world because we are calculating real things and, just as Newton's gravitational constant G carries all the information about the strength of gravity, so g carries all the information about the strength of the electromagnetic force.[4]

If we were actually doing the full calculation, we'd now turn our attention to the second diagram, which represents another way that our original pair of electrons can make their way to the same points, X and Y. The second diagram is very similar to the first in that the electrons start out from the same places, but now the photon is emitted from the upper electron at a different point in space and at a different time and it is absorbed by the lower electron at some other new place and time. Otherwise things run through in precisely the same way and we'll get a second clock, 'clock 2', denoted C2.

Then, on we'd go, repeating the entire process again and again for each and every possible place where the photon can be emitted and each and every possible place where it can be absorbed. We should also account for the fact that the electrons can start out from a variety of different possible starting positions. The key idea is that each and every way of delivering electrons to X and Y needs to be

4. g is related to the fine structure constant: $\alpha = \dfrac{g^2}{4\pi}$

considered, and each is associated with its own clock. Once we have collected together all of the clocks, we 'simply' add them all together, to produce one final clock whose size tells us the probability of finding one electron at X and a second at Y. Then we are finished – we will have figured out how two electrons interact with each other because we can do no better than compute probabilities.

What we have just described really is the heart of QED, and the other forces in Nature admit a satisfyingly similar description. We will come on to those shortly, but first we have a little more to discover.

Firstly, a paragraph describing two small, but important, details. Number 1: we have simplified matters by ignoring the fact that electrons have spin and therefore come in two types. Not only that, photons also have spin (they are bosons) and come in three types. This just makes the calculations a little more messy because we need to keep track of which types of photon and electron we are dealing with at every stage of the hopping and branching. Number 2: if you have been reading carefully then you may have spotted the minus signs in front of a couple of the pictures in Figure 10.1. They are there because we are talking about identical electrons hopping their way to X and Y and the two pictures with the minus sign correspond to an interchange of the electrons relative to the other pictures, which is to say that an electron which started out at one of the upper cluster of points ends up at Y whilst the other, lower, electron ends up at X. And as we argued in Chapter 7, these swapped configurations get combined only after an extra 6-hour wind of their clocks – hence the minus sign.

You may also have spotted a possible flaw in our plan – there are an infinite number of diagrams describing how two electrons can make their way to X and Y, and summing an infinite number of clocks might seem onerous to say the least. Fortunately, every appearance of a photon–electron branching introduces another factor of g into the calculation, and this shrinks the size of the resultant clock. This means that the more complicated the diagram, the smaller the clock it will contribute and the less important it will be when we come to add all the clocks up. For QED, g is quite a small number (it's around 0.3), and so the shrinking is pretty severe as the number of branchings increases. Very often, it is enough to consider only diagrams like the first five in the figure, where there are no more than two branchings, and that saves lots of hard work.

This process of calculating the clock (known in the jargon as the 'amplitude') for each Feynman diagram, adding all the clocks together and squaring the final clock to get a probability that the process will happen is the bread and butter of modern particle physics. But there is a fascinating issue hiding away beneath the surface of all that we have been saying – an issue that bothers some physicists a lot and others not at all.

The Quantum Measurement Problem

When we add the clocks corresponding to the different Feynman diagrams together, we are allowing for the orgy of quantum interference to happen. Just as for the case of the double-slit experiment, where we had to consider every possible route that the particle could take on its journey to the screen, we must consider every possible way that a pair of particles can get from their starting positions to their final positions. This allows us to compute the right answer because it allows for interference between the different diagrams. Only at the end of the process, when all of the clocks have been added together and all the interference is accounted for, should we square up the size of the final clock to calculate the probability that the process will happen. Simple. But look at Figure 10.2.

What happens if we attempt to identify what the electrons are doing as they hop to X and Y? The only way we can examine what is going on is to interact with the system according to the rules of the game. In QED, this means that we must stick to the electron–photon branching rule, because there is nothing else. So let's interact with one of the photons that can be emitted from one or other of the electrons, by detecting it using our own personal photon detector; our eye. Notice that we are now asking a different question of the

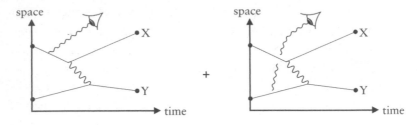

Figure 10.2. A human eye taking a look at what is going on.

theory: 'What is the chance to find an electron at X and another at Y and also a photon in my eye?' We know what to do to get the answer – we should add together all of the clocks associated with the different diagrams that start out with two electrons and end up with an electron at X, another at Y, and also a photon 'in my eye'. More precisely, we should talk about how the photon interacts with my eye. Although that might start out simple enough, it soon gets out of hand. For example, the photon will scatter off an electron sitting in an atom in my eye, and that will trigger a chain of events leading ultimately to my perception of the photon as I become consciously aware of a flash of light in my eye. So to describe fully what is happening involves specifying the positions of every particle in my brain as they respond to the arrival of the photon. We are sailing close to something called the quantum measurement problem.

So far in the book we have described in some detail how to compute probabilities in quantum physics. By that, we mean that quantum theory allows us to calculate the chances of measuring some particular outcome if we conduct an experiment. There is no ambiguity in this process, provided we follow the rules of the game and stick to computing the probabilities of something happening. There is, however, something to feel uneasy about. Imagine an experimenter conducting an experiment for which there are only two outcomes, 'yes' and 'no'. Now imagine actually doing the experiment. The experimenter will record either 'yes' or 'no', and obviously not both at the same time. So far, so good.

Now imagine some future measurement of something else (it

doesn't matter what) made by a second experimenter. Again, we'll assume it is a simple experiment whose outcome is to make a 'click' or 'no click'. The rules of quantum physics dictate that we must compute the probability that the second experiment goes 'click' by summing clocks associated with all of the possibilities that lead to this outcome. Now this may include the circumstance where the first experimenter measures 'yes' *and* the complementary case where they measure 'no'. Only after summing over the two do we get the correct answer for the chances of measuring a 'click' in the second experiment. Is that really right? Do we really have to entertain the notion that, even after the outcome of some measurement, we should maintain the coherence of the world? Or is it the case that once we measure 'yes' or 'no' in the first experiment then the future is dependent only upon that measurement? For example, in our second experiment it would mean that if the first experimenter measures 'yes' then the probability that the second experiment goes 'click' should be computed not from a coherent sum over the 'yes' and 'no' possibilities but instead by considering only the ways in which the world can evolve from 'first experimenter measures yes' to 'second experiment goes click'. This will clearly give a different answer from the case where we are to sum over both the 'yes' and 'no' outcomes and we need to know which is the right thing to do if we are to claim a full understanding.

The way to check which is right is to determine whether there is anything at all special about the measurement process itself. Does it change the world and stop us from adding together quantum amplitudes or rather is measurement part of a vast complex web of possibilities that remain forever in coherent superposition? As human beings we might be tempted to think that measuring something now ('yes' or 'no' say) irrevocably changes the future and if that were true then no future measurement could occur via both the 'yes' and 'no' routes. But it is far from clear that this is the case because it seems that there is always a chance to find the Universe in a future state which can be arrived at via either the 'yes' or 'no' routes. For those states, the laws of quantum physics, taken literally, leave us with no option but to compute the probability of their manifestation by summing over both the 'yes' and 'no' routes. Weird though this may seem, it is no more weird than the summing over histories that we have been performing throughout this book. All that is happening is that we are taking the idea so seriously that we are prepared to do it even at the level of human beings and their actions. From this point of view there is no 'measurement problem'. It is only when we insist that the act of measuring 'yes' or 'no' really changes the nature of things that we run into a problem, because it is then incumbent upon us to explain what it is that triggers the change and breaks the quantum coherence.

The approach to quantum mechanics that we have been discussing, which rejects the idea that Nature goes about choosing a particular version of reality every time someone (or something) 'makes a measurement', forms the basis of what is often referred to as the 'many worlds' interpretation. It is very appealing because it is the logical consequence of taking the laws that govern the behaviour of elementary particles seriously enough to use them to describe all phenomena. But the implications are striking, for we are to imagine that the Universe is really a coherent superposition of all of the possible things that can happen and the world as we perceive it (with its apparently concrete reality) arises only because we are fooled into thinking that coherence is lost every time we 'measure' something. In other words, my conscious perception of the world is fashioned because the alternative (potentially interfering) histories are highly unlikely to lead to the same 'now' and that means quantum interference is negligible.

If measurement is not really destroying quantum coherence then, in a sense, we live out our lives inside one giant Feynman diagram and our predisposition to think that definite things are happening is really a consequence of our crude perceptions of the world. It really is conceivable that, at some time in our future, something can happen to us which requires that, in the past, we did two mutually opposite things. Clearly, the effect is subtle because 'getting the job' and 'not getting the job' makes a big difference to our lives and one cannot easily imagine a scenario where they lead to identical future Universes (remember, we should only add amplitudes that lead to identical outcomes). So in that case, getting and not getting the job do not interfere much with each other and our perception of the world is as if one thing has happened and not the other. However, things become more ambiguous the less dramatic the two alternative scenarios are and, as we have seen, for interactions involving small numbers of particles summing over the different possibilities is absolutely necessary. The large numbers of particles involved in everyday life mean that two substantially different configurations of atoms at some time (e.g. getting the job or not) are simply very unlikely to lead to significantly interfering contributions to some future scenario. In turn, that means we can go ahead and pretend that the world has changed irrevocably as a result of a measurement, even when nothing of the sort has actually happened.

But these musings are not of pressing importance when it comes to the serious business of computing the probability that something will happen when we actually carry out an experiment. For that, we know the rules and we can implement them without any problems. But that happy circumstance may change one day – for now it is the case that questions about how our past might influence the future through quantum interference simply haven't been accessible to experiment. The extent to which meditations on the 'true nature' of the world (or worlds) described by quantum theory can detract from scientific progress is nicely encapsulated in the position taken by the 'shut up and calculate' school of physics, which deftly dismisses any attempt to talk about the reality of things.

Anti-matter

Back in this world, Figure 10.3 shows another way that two electrons can scatter off each other. One of the incoming electrons hops from A to X, whereupon it emits a photon. So far so good but now the electron heads backwards in time to Y where it absorbs another

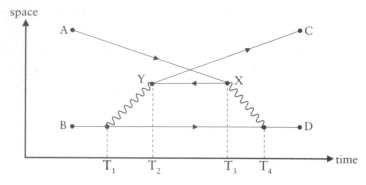

Figure 10.3. Anti-matter . . . or an electron travelling backwards in time.

photon and thence it heads into the future, where it might be eventually detected at C. This diagram does not contravene our rules for hopping and branching, because the electron goes about emitting and absorbing photons as prescribed by the theory. It can happen according to the rules and, as the title of the book suggests, if it can happen, then it does. But such behaviour does appear to violate the rules of common sense, because we are entertaining the idea that electrons travel backwards in time. This would make for nice science fiction, but violating the law of cause and effect is no way to build a universe. It would also seem to place quantum theory in direct conflict with Einstein's Theory of Special Relativity.

Remarkably, this particular kind of time travel for subatomic particles is not forbidden, as Dirac realized in 1928. We can see a hint that all may not be quite as defective as it seems if we reinterpret the goings-on in Figure 10.3 from our 'forwards in time' perspective. We are to track events from left to right in the figure. Let's start at time $T = 0$, where there is a world of just two electrons located at A and B. We continue with a world containing just two electrons until time T_1, whereupon the lower electron emits a photon; between times T_1 and T_2 the world now contains two electrons plus one photon. At time T_2, the photon dies and is replaced by an electron (which will end up at C) and a second particle (which will end up at X). We hesitate to call the second particle an electron because it is 'an electron travelling back in time'. The question is, what does an electron that is travelling back in time look like from the point of view of someone (like you) travelling forwards in time?

To answer this, let's imagine shooting some video footage of an electron as it travels in the vicinity of a magnet, as illustrated in Figure 10.4. Providing that the electron isn't travelling too fast,[5] it will typically travel around in a circle. That electrons can be deflected

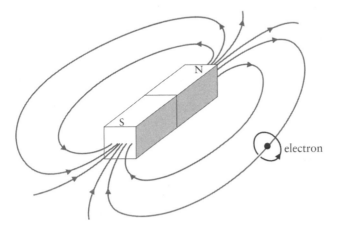

Figure 10.4. An electron, circling near a magnet.

by a magnet is, as we have said before, the basic idea behind the construction of old-fashioned CRT television sets and, more glamorously, particle accelerators, including the Large Hadron Collider. Now imagine that we take the video footage and play it backwards. This is what 'an electron going backwards in time' would look like from our 'forwards in time' perspective. We'd now see the 'backwards in time electron' circle in the opposite direction as the movie advances. From a physicist's perspective, the backwards in time video will look exactly like a forwards in time video shot using a particle which is in every way identical to an electron except that the particle appears to carry positive electric charge. Now we have the answer to our question: electrons travelling backwards in time

5. This is a technical point, to ensure that the electron feels roughly the same sized magnetic force as it moves around.

would appear, to us, as 'electrons of positive charge'. Thus, if electrons do actually travel back in time then we expect to encounter them as 'electrons of positive charge'.

Such particles do exist and they are called 'positrons'. They were introduced by Dirac in early 1931 to solve a problem with his quantum mechanical equation for the electron – namely that the equation appeared to predict the existence of particles with negative energy. Later, Dirac gave a wonderful insight into his way of thinking, and in particular his strong conviction in the correctness of his mathematics: 'I was reconciled to the fact that the negative energy states could not be excluded from the mathematical theory, and so I thought, let us try to find a physical explanation for them.'

Just over a year later, and apparently unaware of Dirac's prediction, Carl Anderson saw some strange tracks in his experimental apparatus while observing cosmic ray particles. His conclusion was that, 'It seems necessary to call upon a positively charged particle having a mass comparable with that of an electron.' Once again, this illustrates the wonderful power of mathematical reasoning. In order to make sense of a piece of mathematics, Dirac introduced the concept of a new particle – the positron – and a few months later it was found, produced in high-energy cosmic ray collisions. The positron is our first encounter with that staple of science fiction, anti-matter.

Armed with this interpretation of time-travelling electrons as positrons, we can finish off the job of explaining Figure 10.3. We are to say that when the photon reaches Y at time T_2 it splits into an electron and a positron. Each head forwards in time until time T_3 when the positron from Y reaches X, whereupon it fuses with the original upper electron to produce a second photon. This photon propagates to time T_4, when it gets absorbed by the lower electron.

This might all sound a little far fetched: anti-particles have emerged from our theory because we are permitting particles to travel backwards in time. Our hopping and branching rules allow particles to hop both forwards and backwards in time, and despite our possible prejudice that this must be disallowed, it turns out that we do not, indeed must not, prevent them from doing so. Quite ironically, it turns out that if we did *not* allow particles to hop back in time then we would have a violation of the law of cause and effect. This is odd, because it seems as if things ought to be the other way around.

That things work out just fine is not an accident and it hints at a deeper mathematical structure. In fact, you may have got the feeling on reading this chapter that the branching and hopping rules all seem rather arbitrary. Could we make up some new branching rules and tweak the hopping rules then explore the consequences? Well, if we did that we would almost certainly build a bad theory – one that would violate the law of cause and effect, for example. Quantum Field Theory (QFT) is the name for the deeper mathematical structure that underpins the hopping and branching rules and it is remarkable for being the *only* way to build a quantum theory of tiny particles that also respects the Theory of Special Relativity. Armed with the apparatus of QFT, the hopping and branching rules are fixed and we lose the freedom to choose. This is a very important result for those in pursuit of fundamental laws because using 'symmetry' to remove choice creates the impression that the Universe simply has to be 'like this' and that feels like progress in understanding. We used the word 'symmetry' here and it is appropriate, because Einstein's theories can be viewed as imposing symmetry restrictions on the structure of space and time. Other 'symmetries' further constrain the hopping and branching rules, and we shall briefly encounter those in the next chapter.

Before leaving QED, we have a final loose end to tie up. If you recall, the opening talk of the Shelter Island meeting concerned the Lamb shift, an anomaly in the hydrogen spectrum that could not be explained by the quantum theory of Heisenberg and Schrödinger. Within a week of the meeting, Hans Bethe produced a first, approximate, calculation of the answer. Figure 10.5 illustrates the QED way to picture a hydrogen atom. The electromagnetic interaction that keeps the proton and the electron bound together can be represented by a series of Feynman diagrams of increasing complexity, just as we saw for the case of two electrons interacting together in

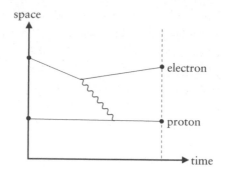

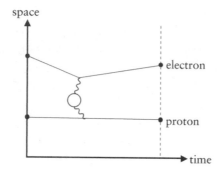

Figure 10.5. The hydrogen atom.

Figure 10.1. We've sketched two of the simplest possible diagrams in Figure 10.5. Pre-QED, the calculations of the electron energy levels included only the top diagram in the figure, which captures the physics of an electron that is trapped within the potential well generated by the proton. But, as we've discovered, there are many other things that can happen during the interaction. The second diagram in Figure 10.5 shows the photon briefly fluctuating into an electron–positron pair, and this process must also be included in a calculation of the possible energy levels of the electron. This, and many other diagrams, enter the calculation as small corrections to the main result.[6]

6. The one first anticipated by Bohr back in 1913.

Bethe correctly included the important effects from 'one-loop' diagrams, like that in the figure, and found that they slightly shift the energy levels and therefore the detail in the observed spectrum of light. His result was in accord with Lamb's measurement. QED, in other words, forces us to imagine a hydrogen atom as a fizzing cacophony of subatomic particles popping in and out of existence. The Lamb shift was humankind's first direct encounter with these ethereal quantum fluctuations.

It did not take long for two other Shelter Island attendees, Richard Feynman and Julian Schwinger, to pick up the baton and, within a couple of years, QED had been developed into the theory we know today – the prototypical quantum field theory and exemplar for the soon-to-be-discovered theories describing the weak and strong interactions. For their efforts, Feynman, Schwinger and the Japanese physicist Sin-Itiro Tomonaga received the 1965 Nobel Prize 'for their fundamental work in quantum electrodynamics, with deep-ploughing consequences for the physics of elementary particles'. It is to those deep-ploughing consequences that we now turn.

11.   Empty Space Isn't Empty

Not everything in the world stems from the interactions between electrically charged particles. QED does not explain the 'strong nuclear' processes that bind quarks together inside protons and neutrons or the 'weak nuclear' processes that keep our Sun burning. We can't write a book about the quantum theory of Nature and leave out half of the fundamental forces, so this chapter will make right our omission before delving into empty space itself. As we'll discover, the vacuum is an interesting place, filled with possibilities and obstacles for particles to navigate.

The first thing to emphasize is that the weak and strong nuclear forces are described by exactly the same quantum field theoretic approach that we have described for QED. It is in this sense that the work of Feynman, Schwinger and Tomonaga had deep-ploughing consequences. Taken as a whole, the theory of these three forces is known, rather unassumingly, as the Standard Model of particle physics. As we write, the Standard Model is being tested to breaking point by the largest and most sophisticated machine ever assembled: CERN's Large Hadron Collider (LHC). 'Breaking point' is right because, in the absence of something hitherto undiscovered, the Standard Model stops making meaningful predictions at the energies involved in the collisions of almost light-speed protons at the LHC. In the language of this book, the quantum rules start to generate clock faces with hands longer than 1, which means that certain processes involving the weak nuclear force are predicted to occur with a probability greater than 100%. This is clearly nonsense and it implies that the LHC is destined to discover something new. The challenge is to identify it among the hundreds of millions of proton collisions generated every second a hundred metres below the foothills of the Jura Mountains.

The Standard Model does contain a cure to the malaise of the dysfunctional probabilities and that goes by the name of the 'Higgs mechanism', which predicts that the LHC should observe one more particle of Nature, the Higgs boson, and with it trigger a profound shift in our view of what constitutes empty space. We'll get to the Higgs mechanism later in the chapter, but first we should provide a short introduction to the triumphant yet creaking Standard Model.

The Standard Model of Particle Physics

In Figure 11.1 we've listed all of the known particles. These are the building blocks of our Universe, but we expect that there may be some more – perhaps we will see a new particle associated with the abundant but enigmatic Dark Matter that seems necessary to explain the Universe at large. Or perhaps the supersymmetric particles anticipated by string theory or maybe the Kaluza-Klein excitations characteristic of extra dimensions in space or techniquarks or lepto-quarks or . . . theoretical speculation is rife and it is the duty of those carrying out experiments at the LHC to narrow down the field, rule out the wrong theories and point the way forward.

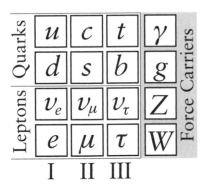

Figure 11.1. The particles of Nature.

Everything you can see and touch; every inanimate machine, every living thing, every rock and every human being on planet Earth, every planet and every star in every one of the 350 billion galaxies in the observable Universe is built out of the particles in the first column of four. You are an arrangement of just three: the up and down quarks and the electron. The quarks make up your atomic nuclei and, as we've seen, the electrons do the chemistry. The remaining particle in the first column, called the electron neutrino, may be less familiar to you but there are around 60 billion of them streaming through every square centimetre of your body every second from the Sun. They mostly sail straight through you and the entire Earth, unimpeded, which is why you've never seen or felt one. But they do, as we will see in a moment, play a crucial role in the processes that power the Sun and, because of that, they make your life possible.

These four particles form a set known as the first generation of matter and, together with the four fundamental forces of Nature, they appear to be all that is needed to build a Universe. For reasons that we do not yet understand, Nature has chosen to provide us with two further generations – clones of the first except that the particles are more massive. They are represented in the second and third columns in Figure 11.1. The top quark in particular is much more massive than the other fundamental particles. It was discovered at the Tevatron accelerator at Fermilab near Chicago in 1995, and its mass has been measured to be over 180 times the mass of a proton. Why the top quark is such a monster, while being point-like in the same way that an electron is point-like, is a mystery. Although these extra generations of matter do not play a direct role in the ordinary affairs of the Universe they do seem to have been crucial players in the moments just after the Big Bang . . . but that is another story.

Also shown in Figure 11.1, in the column on the right, are the force-carrying particles. Gravity is not represented in the table because we do not have a quantum theory of gravity that sits comfortably within the framework of the Standard Model. This isn't to say that there isn't one; string theory is an attempt to bring gravity into the fold but, to date, it has met with limited success. Because gravity is so feeble it plays no significant role in particle physics experiments and for that pragmatic reason we'll say no more about it. We learnt in the last chapter how the photon is responsible for mediating the electromagnetic force between electrically charged particles and that its behaviour was determined by specifying a new branching rule. The W and Z particles do the corresponding job for the weak force while the gluons mediate the strong force. The primary differences between the quantum descriptions of the forces arise because the branching rules are different. It is (almost) that simple and we have drawn some of the new branching rules in Figure 11.2. The similarity with QED makes it easy to appreciate the basics of the weak and strong forces; we just need to know what the branching rules are and then we can draw Feynman diagrams like we did for QED in the last chapter. Fortunately, changing the branching rules makes all the difference to the physical world.

If this were a particle physics textbook, we might proceed to outline the branching rules for each of the processes in Figure 11.2, and many more besides. These rules, known as the Feynman rules, would then allow you, or a computer program, to calculate the probability for some process or other, just as we outlined in the last chapter for QED. The rules capture something essential about the world and it is delightful that they can be summarized in a few simple pictures and rules. But this isn't a particle physics textbook, so we'll instead focus on the top-right diagram, because it is a particularly important branching rule for life on Earth. It shows an up quark branching into a down quark by emitting a W particle and this behaviour is exploited to dramatic effect within the core of the Sun.

The Sun is a gaseous sea of protons, neutrons, electrons and photons with the volume of a million earths, collapsing under its own gravity. The vicious compression heats the solar core to 15 million degrees and at these temperatures the protons begin to fuse together to form helium nuclei. The fusion process releases energy, which increases the pressure on the outer layers of the star, balancing the

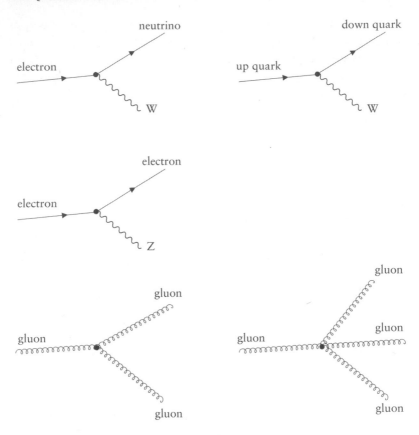

Figure 11.2. Some of the branching rules for the weak and strong forces.

inward pull of gravity. We'll dig deeper into this precarious balancing act in the epilogue, but for now we want to understand what it means to say that 'the protons begin to fuse together'.

This sounds simple enough, but the precise mechanism for fusion in the Sun's core was a source of great scientific debate during the 1920s and 30s. The British scientist Arthur Eddington was the first to propose that the energy source of the Sun is nuclear fusion, but it was quickly pointed out that the temperatures were apparently far too low for the process to occur given the then-known laws of physics. Eddington stuck to his guns, however, issuing the famous retort: 'The helium which we handle must have been put together at some time and some place. We do not argue with the critic who urges that the stars are not hot enough for this process; we tell him to go and find a hotter place.'

The problem is that when two fast-moving protons in the core of the Sun get close, they repel each other as a result of the electromagnetic force (or, in the language of QED, by photon exchange). To fuse together they need to get so close that they are effectively overlapping and, as Eddington and his colleagues well knew, the solar protons are not moving fast enough (because the Sun is not hot enough) to overcome their mutual electromagnetic repulsion.

The answer to this conundrum is that the W particle steps in to save the day. In a stroke, one of the protons in the collision can convert into a neutron by converting one of its up quarks into a down quark, as specified by the branching rule in Figure 11.2. Now the newly formed neutron and remaining proton can get very close, because the neutron carries no electric charge. In the language of quantum field theory, this means there is no photon exchange to push the neutron and proton apart. Freed from the electromagnetic repulsion, the proton and neutron can fuse together (as a result of the strong force) to make a deuteron and this quickly leads to helium formation, releasing life-giving energy for the star. The process is illustrated in Figure 11.3, which also indicates that the W particle does not stick around for very long; instead it branches into a positron and a neutrino – this is the source of those very same neutrinos that pass through your body in such vast numbers. Eddington's belligerent defence of fusion as the power source of the Sun was correct, although he could have had no inkling of the solution. The all-important W particle, along with its partner the Z, was eventually discovered at CERN in the 1980s.

To conclude our brief survey of the Standard Model, we turn to the strong force. The branching rules are such that only quarks can branch into gluons. In fact they are much more likely to do that than they are

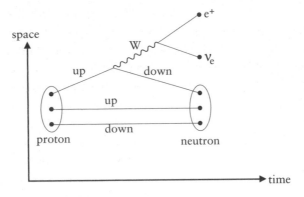

Figure 11.3. Proton conversion into a neutron by weak decay, with the emission of a positron and a neutrino. Without this, the Sun would not burn.

to do anything else. This predisposition to emit gluons is why the strong force is so named and it is the reason why gluon branching is able to defeat the repulsive electromagnetic force that would otherwise cause the positively charged proton to explode. Fortunately, the strong force cannot reach very far. Gluons tend not to travel beyond around 1 femtometre (10^{-15} m) before they branch again. The reason why gluons are so short-ranging in their influence, whilst photons can reach across the Universe, is down to the fact that gluons can also branch into other gluons, as illustrated in the final two pictures in Figure 11.2. This trick of the gluons makes the strong force very different from the electromagnetic force, and effectively confines its actions to the interior of the atomic nucleus. Photons have no such self-branching and that is very fortunate, for if they did you wouldn't be able to see the world in front of your eyes because the photons streaming towards you would scatter off those travelling across your line of sight. It is one of the wonders of life that we can see anything at all, and a vivid reminder that photons very rarely interact with each other.

We have not explained where all of these new rules come from, nor have we explained why the Universe contains the particles that it does. There is a good reason for this: we don't really know the answers to either of these questions. The particles that make up our

Universe – the electrons, neutrinos and quarks – are the primary actors in the unfolding cosmic drama, but to date we have no compelling way to explain why the cast should line up as it does.

What is true, however, is that once we have the list of particles then the way they interact with each other, as prescribed by the branching rules, is something we can partially anticipate. The branching rules are not something that physicists have just conjured from nowhere – they are in all cases anticipated on the grounds that the theory describing the particle interactions should be a Quantum Field Theory supplemented with something called gauge symmetry. To discuss the origin of the branching rules would take us too far outside the main line of this book – but we do want to reiterate that the essential rules are very simple: the Universe is built from particles that move around and interact according to a handful of hopping and branching rules. We can take those rules and use them to compute the probability that 'something' *does* happen by adding together a bunch of clocks – there being one clock for each and every way that the 'something' *can* happen.

The Origin of Mass

By introducing the idea that particles can branch as well as hop we have entered into the domain of Quantum Field Theory, and hopping and branching is, to a large extent, all there is to it. We have, however, been rather negligent in our discussion of mass, for the good reason that we have been saving the best until last.

Modern-day particle physics aims to provide an answer to the question 'what is the origin of mass?' and it does so with the help of a beautiful and subtle piece of physics and a new particle – new in the sense that we have not yet really encountered it in this book, and new in the sense that people on Earth probably realized that they encountered one 'face to face' for the first time in 2012. The particle is named the Higgs boson. When we first wrote this book in September 2011, there were tantalizing glimpses of a Higgs-like object in the LHC data, but there are simply not enough events[1] to decide one way or the other. We say that, 'It may well be that, as you read this book, the situation has changed and the Higgs is a reality. Or it may be that the interesting signals have vanished under further scrutiny.' Well, the situation has changed and the evidence has firmed up: a new particle has been discovered at CERN and it has the hallmark of being Higgs boson. The particularly exciting thing about the question of the origin of mass is that the answer is extremely interesting beyond the obvious desire to know what mass is. Let us now explain that rather cryptic and offensively constructed sentence in more detail.

When we discussed photons and electrons in QED, we introduced the hopping rule for each and said that they are different – we used the symbol P(A,B) for the rule associated with an electron that hops from A to B and the symbol L(A,B) for the corresponding rule for a photon. It is time now to investigate why the rule is different in the two cases. There is a difference because electrons come in two different types (as we know, they 'spin' in one of two different ways), whilst photons come in three different types, but that particular difference will not concern us here. There is another difference, however, because the electron has mass while the photon does not – this is what we want to explore.

Figure 11.4 illustrates one way that we are allowed to think about the propagation of a massive particle. The figure shows a particle hopping from A to B in stages. It goes from A to point 1, from point 1 to point 2 and so on until it finally hops from point 6 to B. What is interesting is that, when written in this way, the rule for each hop is the rule for a particle with *zero* mass, but with one important caveat: every time the particle changes direction we are to apply a new shrinking rule, with the amount of shrinking inversely proportional

---

1. An 'event' is a single proton–proton collision. Because fundamental physics is a counting game (it works with probabilities) it is necessary to keep colliding protons in order to accumulate a sufficient number of those very rare events in which a Higgs particle is produced. What constitutes a sufficient number depends on how skilful the experimenters are at confidently eliminating fake signals.

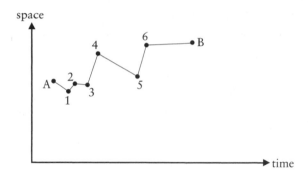

Figure 11.4. A massive particle travelling from A to B.

to the mass of the particle we are describing. This means that, at each kink, the clocks of heavy particles receive less shrinking than the clocks of lighter particles. It is important to emphasize that this isn't an ad hoc prescription. Both the zig-zag and the shrink emerge directly from the Feynman rules for the propagation of a massive particle, without any further assumptions.[2] Figure 11.4 shows just one way that our heavy particle can get from A to B, i.e. via six kinks and six shrinkage factors. To get the final clock associated with a massive particle hopping from A to B we must, as always, add together the infinity of clocks associated with all of the possible ways that the particle can zig-zag its way from A to B. The simplest route is the direct one, with no kinks, but routes with huge numbers of kinks need to be considered too.

For particles with zero mass the shrinkage factor associated with each kink is a killer, because it is infinite. In other words, we are to shrink the clock to zero after the first kink. The only route that matters for massless particles is therefore the direct route – there is simply no clock associated with any other route. This is exactly what we would expect: it means that we can use the hopping rule for massless

---

2. Our ability to think of a massive particle as a massless particle supplemented with a 'kink' rule comes from the fact that $P(A,B) = L(A,B) + L(A,1)L(1,B)S + L(A,1)L(1,2)L(2,B)S^2 + L(A,1)L(1,2)L(2,3)L(3,B)S^3 + \ldots$, where S is the shrinkage factor associated with a kink and it is understood that we should sum over all possible intermediate points 1, 2, 3 etc.

particles when the particle is massless. However, for particles with non-zero mass, kinks are allowed, although if the particle is very light then the shrinking factor imposes a severe penalty on paths with many kinks. The most likely paths are therefore those with very few kinks. Conversely, heavy particles do not get penalized much when they kink, and so they tend to be described by paths with lots of zig-zagging. This seems to suggest that heavy particles really ought to be thought of as massless particles that zig-zag their way from A to B. The amount of zig-zagging is what we identify as 'mass'.

This is all rather nice, for we have a new way to think about massive particles. Figure 11.5 illustrates the propagation from A to B of three different particles of increasing mass. In each case, the rule associated with each 'zig' or 'zag' of the path is the same as that for a massless particle, and for every kink we are to pay a 'the clock must be shrunk' penalty. We should not get overly excited yet because we have not really explained anything fundamental. All we have done is to replace the word 'mass' with the words 'tendency to zig-zag'. We are allowed to do this because they are mathematically equivalent descriptions of the propagation of a massive particle. But even so, it feels like an interesting thing and, as we shall now discover, it may turn out to be rather more than just a mathematical curiosity.

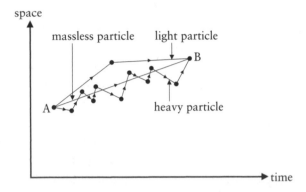

Figure 11.5. Particles of increasing mass propagating from A to B. The more massive a particle is the more it zig-zags.

In 2012, the LHC was busy colliding protons together with a combined energy of 8 TeV. 'TeV' stands for Tera electron volts, which corresponds to the amount of energy an electron would have if it were accelerated through a potential difference of 8 million million volts. To get a sense of how much energy this is, it's roughly the energy that subatomic particles would have had about a trillionth of a second after the Big Bang and it is enough energy to conjure out of thin air a mass equal to 8,000 protons (via Einstein's $E = mc^2$). And this is only half the design energy; the LHC has more gas in the tank.

One of the primary reasons that eighty-five countries around the world have come together to build and operate this vast, audacious experiment is to hunt for the mechanism that is responsible for generating the masses of the fundamental particles. The most widely accepted theory for the origin of mass works by providing an explanation for the zig-zagging: it posits a new fundamental particle that the other particles 'bump into' on their way through the Universe.

That particle is the Higgs boson. According to the Standard Model, without a Higgs the fundamental particles would hop from place to place without any zig-zagging and the Universe would be a very different place. But if we fill empty space with Higgs particles then they can act to deflect particles, making them zig-zag and, as we have just learnt, that leads to the emergence of 'mass'. It is rather like trying to walk through a crowded pub – one gets buffeted from side-to-side and ends up taking a zig-zag path towards the bar.

The Higgs mechanism is named after Edinburgh theorist Peter Higgs and it was introduced into particle physics in 1964. The idea was obviously very ripe because several people came up with the idea at the same time – Higgs of course, and also Robert Brout and François Englert working in Brussels and Gerald Guralnik, Carl Hagan and Tom Kibble in London. Their work was itself built on the earlier efforts of many others, including Heisenberg, Yoichiro Nambu, Jeffrey Goldstone, Philip Anderson and Weinberg. The full realization of the idea, for which Sheldon Glashow, Abdus Salam and Weinberg received the Nobel Prize in 1979, is no less than the

Standard Model of particle physics. The idea is simple enough – empty space is not empty, and this leads to zig-zagging and therefore mass. But clearly we have some more explaining to do. How can it be that empty space is jammed full of Higgs particles – wouldn't we notice this in our everyday lives, and how did this strange state of affairs come about in the first place? It certainly sounds like a rather extravagant proposition. We have also not explained how it can be that some particles (like photons) have no mass while others (like W bosons and top quarks) weigh in with masses comparable to that of an atom of silver or gold.

The second question is easier to answer than the first, at least superficially. Particles only ever interact with each other through a branching rule and Higgs particles are no different in that regard. The branching rule for a top quark includes the possibility that it can couple to a Higgs particle, and the corresponding shrinking of the clock (remember all branching rules come with a shrinking factor) is much less than it is in the case of the lighter quarks. That is 'why' a top quark is so much heavier than an up quark. This doesn't explain why the branching rule is what it is, of course. The current answer to that is the disappointing 'because it is'. It's on the same footing as the question 'Why are there three generations of particles?' or 'Why is gravity so weak?' Similarly, photons do not have any branching rule that couples them to Higgs particles and as a result they do not interact with them. This, in turn, means that they do not zig-zag and have no mass. Although we have passed the buck to some extent, this does feel like some kind of an explanation, and it is certainly true that by producing Higgs particles at the LHC it is possible to check that they couple to the other particles in this manner.[3] Then we can legitimately claim to have gained a rather thrilling insight into the way Nature works.[3]

The first of our outstanding questions is a little trickier to explain – namely, how can it be that empty space is full of Higgs particles? To get warmed up, we need to be very clear about one

---

3. At the time of this writing, the data are not inconsistent with this but further analysis is needed before the couplings will be determined.

thing: quantum physics implies that there is no such thing as empty space. In fact, what we call 'empty space' is really a seething maelstrom of subatomic particles and there is no way to sweep them away and clean it up. Once we realize that, it becomes much less of an intellectual challenge to accept that empty space might be full of Higgs particles. But let's take one step at a time.

You might imagine a tiny region of deep outer space, a lonely corner of the Universe millions of light years from a galaxy. As time passes it is impossible to prevent particles from appearing and then disappearing out of nothing. Why? It is because the process of the creation and annihilation of particle–anti-particle pairs is allowed by the rules. An example can be found in the lower diagram in Figure 10.5: imagine stripping away everything except for the electron loop – the diagram then corresponds to an electron–positron pair spontaneously appearing from nothing and then disappearing back into nothing. Because drawing a loop does not violate any of the rules of QED we must acknowledge that it is a real possibility; remember, everything that can happen does happen. This particular possibility is just one of an infinite number of ways that empty space can fizz and pop, and because we live in a quantum universe the correct thing to do is to add all the possibilities together. The vacuum, in other words, has an incredibly rich structure, made up out of all the possible ways that particles can pop in and out of existence.

That last paragraph introduced the idea that the vacuum is not empty, but we painted a rather democratic picture in which all of the elementary particles play a role. What is it about the Higgs particle that makes it special? If the vacuum were nothing other than a seething broth of matter–antimatter creation and annihilation, then all of the elementary particles would continue to have zero mass – the quantum loops themselves are not capable of delivering it.[4] Instead, we need to populate the vacuum with something different, and this is where the bath of Higgs particles enters. Peter Higgs

4. This is a subtle point and derives from the 'gauge symmetry', which underwrites the hopping and branching rules of the elementary particles.

simply stipulated that empty space is packed with Higgs particles[5] and didn't feel obliged to offer any deep explanation as to why. The Higgs particles in the vacuum provide the zig-zag mechanism and they are working overtime by interacting with each and every massive particle in the Universe, selectively retarding their motion to create mass. The net result of the interactions between ordinary matter and a vacuum full of Higgs particles is that the world goes from being a structureless place to a diverse and wonderful living world of stars, galaxies and people.

The big question of course is where those Higgs particles came from in the first place? The answer isn't really known, but it is thought that they are the remnants of what is known as a phase transition that occurred sometime shortly after the Big Bang. If you are patient and watch the glass in your window as the temperature falls on a winter's evening, you'll see the structured beauty of ice crystals emerge as if by magic from the water vapour in the night air. The transition from water vapour to ice on cold glass is a phase transition – water molecules rearranging themselves into ice crystals; the spontaneous breaking of the symmetry of a formless vapour cloud triggered by a drop in temperature. Ice crystals form because it is energetically more favourable to do so. Just as a ball rolls down the side of a mountain to take up a lower energy in a valley, or electrons rearrange themselves around atomic nuclei to form the bonds that hold molecules together, so the sculpted beauty of a snowflake is a lower energy configuration of water molecules than a formless cloud of vapour.

We think that a similar thing happened early on in the Universe's history. As the hot gas of particles that was the nascent Universe expanded and cooled, so it transpired that a Higgs-free vacuum was energetically disfavoured and a vacuum filled with Higgs particles was the natural state. The process really is similar to the way that water condenses into droplets or ice forms on a cold pane of glass. The spontaneous appearance of water droplets when they con-

5. He was far too modest to call them by that name.

dense on a pane of glass creates the impression that those droplets simply emerged out of 'nothing'. Similarly for the Higgs, in the hot stages just after the Big Bang the vacuum is seething with the fleeting quantum fluctuations (those loops in our Feynman diagrams), as particles and anti-particles pop out of nothing before disappearing again. However, something radical happens as the Universe cools and suddenly, out of nothing, just as the water drops appear on the glass, a 'condensate' of Higgs particles emerges, all held together by their mutual interactions in an ephemeral suspension through which the other particles propagate.

The idea that the vacuum is filled with material suggests that we, and everything else in the Universe, live out our lives inside a giant condensate that emerged as the Universe cooled down, just as the morning dew emerges with the dawn. Lest we think that the vacuum is populated merely as a result of Higgs particle condensation, we should also remark that there is even more to the vacuum than this. As the Universe cooled still further, quarks and gluons also condensed to produce what are, naturally enough, known as quark and gluon condensates. The existence of these is well established by experiments, and they play a very important role in our understanding of the strong nuclear force. In fact, it is this condensation that gives rise to the vast majority of the mass of protons and neutrons. The Higgs vacuum is, however, responsible for generating the observed masses for the elementary particles – the quarks, electrons, muons, taus and W and Z particles. The quark condensate kicks in to explain what happens when a cluster of quarks binds together to make a proton or a neutron. Interestingly, whilst the Higgs mechanism is relatively unimportant when it comes to explaining the mass of protons, neutrons and the heavier atomic nuclei, the converse is true when it comes to explaining the mass of the W and Z particles. For them, quark and gluon condensation would generate a mass of around 1 GeV in the absence of a Higgs particle, but their experimentally measured masses are closer to 100 times this. The LHC was designed to operate in the energy domain of the W and Z, where it can explore the mechanism responsible for their comparatively large masses.

To put some rather surprising numbers on all of this, the energy stored up within 1 cubic metre of empty space as a result of quark and gluon condensation is a staggering 10^{35} joules, and the energy due to Higgs condensation is 100 times larger than this. Together, that's the total amount of energy our Sun produces in 1,000 years. To be precise, this is 'negative' energy, because the vacuum is lower in energy than a Universe containing no particles at all. The negative energy arises because of the binding energy associated with the formation of the condensates, and is not by itself mysterious. It is no more glamorous than the fact that, in order to boil water (and reverse the phase transition from vapour to liquid), you have to put energy in.

What is mysterious, however, is that such a large and negative energy density in every square metre of empty space should, if taken at face value, generate a devastating expansion of the Universe such that no stars or people would ever form. The Universe would literally have blown itself apart moments after the Big Bang. This is what happens if we take the predictions for vacuum condensation from particle physics and plug them directly into Einstein's equations for gravity, applied to the Universe at large. This heinous conundrum goes by the name of the cosmological constant problem and it remains one of the central problems in fundamental physics. Certainly it suggests that we should be very careful before claiming to really understand the nature of the vacuum and / or gravity. There is something absolutely fundamental that we do not yet understand.

With that sentence, we come to the end of our story because we've reached the edge of our knowledge. The domain of the known is not the arena of the research scientist. Quantum theory, as we observed at the beginning of this book, has a reputation for difficulty and downright contrary weirdness, exerting as it does a rather liberal grip on the behaviour of the particles of matter. But everything we've described, with the exception of this final chapter, is known and well understood. Following evidence rather than common sense, we are led to a theory that is manifestly able to describe a vast range of phenomena, from the sharp rainbows emitted by hot atoms to fusion within stars. Putting the theory to use led to the most important technological breakthrough of the twentieth century – the transistor – a device whose operation would be inexplicable without a quantum view of the world.

But quantum theory is far more than a mere explanatory triumph. In the forced marriage between quantum theory and relativity, anti-matter emerged as a theoretical necessity and was duly discovered. Spin, the fundamental property of subatomic particles that underpins the stability of atoms, was likewise a theoretical prediction required for the consistency of the theory. And now, in the second quantum century, the Large Hadron Collider voyages into the unknown to explore the vacuum itself. This is scientific progress; the gradual and careful construction of a legacy of explanation and prediction that changes the way we live. And this is what sets science apart from everything else. It isn't simply another point of view – it reveals a reality that would be impossible to imagine, even for the possessor of the most tortured and surreal imagination. Science is the investigation of the real, and if the real seems surreal then so be it. There is no better demonstration of the power of the scientific method than quantum theory. Nobody could have come up with it without the most meticulous and detailed experiments, and the theoretical physicists who built it were able to suspend and jettison their deeply held and comforting beliefs in order to explain the evidence before them. Perhaps the conundrum of the vacuum energy signals a new quantum journey, perhaps the LHC will provide new and inexplicable data, and perhaps everything in this book will turn out to be an approximation to a much deeper picture – the exciting journey to understand our Quantum Universe continues.

When we began thinking about writing this book, we spent some time debating how to end it. We wanted to find a demonstration of the intellectual and practical power of quantum theory that would convince even the most sceptical reader that science really does describe, in exquisite detail, the workings of the world. We both agreed that there is such a demonstration, although it does involve some algebra – we have done our best to make it possible to follow the reasoning without scrutinizing the equations, but it does come with that warning. So, our book ends here, unless you want a little bit more: the most spectacular demonstration, we think, of the power of quantum theory. Good luck, and enjoy the ride.

Epilogue: the Death of Stars

When stars die, many end up as super-dense balls of nuclear matter intermingled with a sea of electrons, known as 'white dwarves'. This will be the fate of our Sun when it runs out of nuclear fuel in around 5 billion years time. It will also be the fate of over 95% of the stars in our galaxy. Using nothing more than a pen, paper and a little thought, we can calculate the largest possible mass of these stars. The calculation, first performed by Subrahmanyan Chandrasekhar in 1930, uses quantum theory and relativity to make two very clear predictions. Firstly, that there should even be such a thing as a white dwarf star – a ball of matter held up against the crushing force of its own gravity by the Pauli Exclusion Principle. Secondly, that if we turn our attention from the piece of paper with our theoretical scribbles on it and gaze into the night sky then we should *never* see a white dwarf with a mass greater than 1.4 times the mass of our Sun. These are spectacularly audacious predictions.

Today, astronomers have catalogued around 10,000 white dwarf stars. The majority have masses around 0.6 solar masses, but the largest recorded mass is *just* under 1.4 solar masses. This single number, '1.4', is a triumph of the scientific method. It relies on an understanding of nuclear physics, of quantum physics and of Einstein's Theory of Special Relativity – an interlocking swathe of twentieth-century physics. Calculating it also requires the fundamental constants of Nature we've met in this book. By the end of this chapter, we will learn that the maximum mass is determined by the ratio

$$\left(\frac{hc}{G}\right)^{3/2} \frac{1}{m_p^2}$$

Look carefully at what we just wrote down: it depends on Planck's constant, the speed of light, Newton's gravitational constant and the mass of a proton. How wonderful it is that we should be able to predict the uppermost mass of a dying star using this combination of fundamental constants. The three-way combination of gravity, relativity and the quantum of action appearing in the ratio $(hc/G)^{1/2}$ is called the Planck mass, and when we put the numbers in it works out at approximately 55 micrograms; roughly the mass of a grain of sand. So the Chandrasekhar mass is, rather astonishingly, obtained by contemplating two masses, one the size of a grain of sand and the other the mass of a single proton. From such tiny numbers emerges a new fundamental mass scale in Nature: the mass of a dying star.

We could present a very broad overview of how the Chandrasekhar mass comes about, but instead we'd like to do a little bit more: we'd like to describe the actual calculation because that is what really makes the spine tingle. We'll fall short of actually computing the precise number (1.4 solar masses), but we will get close to it and see how professional physicists go about drawing profound conclusions using a sequence of carefully developed logical steps, invoking well-known physical principles along the way. There will be no leap of faith. Instead, we will keep a cool head and slowly and inexorably be drawn to the most exciting of conclusions.

Our starting point has to be: 'what is a star?' The visible Universe is, to a very good approximation, made up of hydrogen and helium, the two simplest elements formed in the first few minutes after the Big Bang. After around half a billion years of expansion, the Universe was cool enough for slightly denser regions in the gas clouds to start clumping together under their own gravity. These were the seeds of the galaxies, and within them, around smaller clumps, the first stars began to form.

The gas in these first proto-stars became hotter and hotter as they collapsed in on themselves, as anyone who has used a bicycle pump will know, because compressing a gas makes it heat up. When the gas reaches temperatures of around 100,000 degrees, the electrons can no longer be held in orbit around the hydrogen and helium nuclei and the atoms get ripped apart, leaving a hot plasma of bare nuclei and electrons. The hot gas tries to expand outwards and resist further collapse but, for sufficiently massive clumps, gravity wins out. Because protons have positive electric charge they will repel each other but, as the gravitational collapse proceeds and the temperature continues to rise, the protons move faster and faster. Eventually, at a temperature of several million degrees, the protons are moving so fast that they get close enough for the weak nuclear force to take over. When that happens, two protons can react with one another; one of them spontaneously changes into a neutron with the simultaneous emission of a positron and a neutrino (exactly as illustrated in Figure 11.3 on page 202). Freed from the electrical repulsion, the proton and the neutron fuse under the action of the strong nuclear force to make a deuteron. This process releases huge amounts of energy because, just as in the formation of a hydrogen molecule, binding things together releases energy.

The energy release in a single fusion event isn't large by everyday standards. One million proton–proton fusion reactions generate roughly the same amount of energy as the kinetic energy of a mosquito in flight or a 100 watt light-bulb radiates in a nanosecond. But that is huge on atomic scales and, remember, we are talking about the dense heart of a collapsing gas cloud in which there are around 10^{26} protons per cubic centimetre. If all the protons in a cubic centimetre were to fuse into deuterons, 10^{13} joules of energy would be liberated, which is enough to power a small town for one year.

The fusion of two protons into a deuteron is the start of a fusion jamboree. The deuteron itself is eager to fuse with a third proton to make a light version of helium (called helium-3) with the emission of a photon, and those helium nuclei then pair up and fuse into regular helium (called helium-4) with the emission of two protons. At each stage, the fusing together liberates more and more energy. And, just for good measure, the positron, which was emitted right back at the start of the chain, also rapidly fuses with an electron in the surrounding plasma to produce a pair of photons. All of this liberated energy makes for a hot gas of photons, electrons and nuclei that pushes against the in-falling matter and halts any further gravitational collapse. This is a star: nuclear fusion burns up nuclear fuel in the core, and that generates an outward pressure that stabilizes the star against gravitational collapse.

There is, of course, only a finite amount of hydrogen fuel available to burn and, eventually, it will run out. With no more energy released there is no more outward pressure; gravity once again takes control and the star resumes its postponed collapse. If the star is massive enough, the core will heat up to temperatures of around 100 million degrees. At that stage, the helium produced as waste in the hydrogen-burning phase ignites, fusing together to produce carbon and oxygen, and once again the gravitational collapse is temporarily halted.

But what happens if the star is not massive enough to initiate helium fusion? For stars less than about half the mass of our Sun, this is the case, and for them something very dramatic happens. The star heats up as it contracts, but, before the core reaches 100 million degrees, something else halts the collapse. That something is the pressure exerted by electrons due to the fact that they are in the grip of the Pauli Exclusion Principle. As we have learnt, the Pauli principle is crucial to understanding how atoms remain stable, and it underpins the properties of matter. Here is another string to its bow: it explains the existence of compact stars that survive despite the fact that they no longer burn up any nuclear fuel. How does this work?

As the star gets squashed, so the electrons within it get confined to a smaller volume. We can think of an electron in the star in terms of its momentum p and hence its associated de Broglie wavelength, h/p. In particular, the particle can only ever be described by a wave packet that is at least as big as its associated wavelength.[1] This means that, when the star is dense enough, the electrons must be overlapping each other, i.e. we cannot imagine them as being described by

---

1. Recall from Chapter 5 that particles of definite momentum are in fact described by infinitely long waves and that as we allow for some spread in the momentum so we can start to localize the particle. But this can only go so far and it makes no sense to talk about a particle of a certain wavelength if it is localized to a distance smaller than that wavelength.

isolated wave packets. This in turn means that quantum mechanical effects, and the Pauli principle in particular, are important in describing the electrons. Specifically, they are being squashed together to the point where two electrons are attempting to occupy the same region of space, and we know from the Pauli principle that they resist this. In a dying star, therefore, the electrons avoid each other and this provides a rigidity that resists any further gravitational collapse.

This is the fate of the lightest stars, but what of stars like our Sun? We left them a couple of paragraphs ago, burning helium into carbon and oxygen. What happens when they run out of helium? They too must then start to collapse under their own gravity, which means they will have their electrons squashed together. And, just as for the lighter stars, the Pauli principle can eventually kick in and halt the collapse. But, for the most massive of stars, even the Pauli Exclusion Principle has its limits. As the star collapses and the electrons get squashed closer together, so the core heats up and the electrons move faster. For heavy enough stars, the electrons will eventually be moving so fast that they approach the speed of light, and that is when something new happens. When they close in on light-speed, the pressure the electrons are able to exert to resist gravity is reduced to such an extent that they aren't up to the job. They simply cannot beat gravity any more and halt the collapse. Our task in this chapter is to calculate when this happens, and we've already given away the punchline. For stars with masses greater than 1.4 times the mass of the Sun, the electrons lose and gravity wins.

That completes the overview that will provide the basis for our calculation. We can now go ahead and forget all about nuclear fusion, because stars that are burning are not where our interest lies. Rather, we are keen to understand what happens inside dead stars. We want to see just how the quantum pressure from the squashed electrons balances the force of gravity, and how that pressure becomes diminished if the electrons are moving too fast. The heart of our study is therefore a balancing game: gravity versus quantum pressure. If we can make them balance we have a white dwarf star, but if gravity wins we have catastrophe.

Although not relevant for our calculation, we can't leave things on such a cliff-hanger. As a massive star implodes, two further options remain open to it. If it is not too heavy then it will keep squashing the protons and electrons until they too can fuse together to make neutrons. In particular, one proton and one electron convert spontaneously into a neutron with the emission of a neutrino, again via the weak nuclear force. In this way the star relentlessly converts into a tiny ball of neutrons. In the words of Russian physicist Lev Landau, the star converts into 'one gigantic nucleus'. Landau wrote those words in his 1932 work 'On the Theory of Stars', which appeared in print in the very same month that the neutron was discovered by James Chadwick. It is probably going too far to say that Landau predicted the existence of neutron stars but, with great prescience, he certainly anticipated something like them. Perhaps the credit should go to Walter Baade and Fritz Zwicky, who wrote in the following year: 'With all reserve we advance the view that supernovae represent the transitions from ordinary stars into neutron stars, which in their final stages consist of extremely closely

Figure 12.1. A cartoon from the 19 January 1934 edition of the *Los Angeles Times*.

packed neutrons.' The idea was considered so outlandish that it was parodied in the *Los Angeles Times* (see Figure 12.1), and neutron stars remained a theoretical curiosity until the mid 1960s.

In 1965, Anthony Hewish and Samuel Okoye found 'evidence for an unusual source of high radio brightness temperature in the crab nebula', although they failed to identify it as a neutron star. The positive ID came in 1967 by Iosif Shklovsky and, shortly afterwards, after more detailed measurements, by Jocelyn Bell and Hewish himself. This first example of one of the most exotic objects in the Universe was subsequently named the 'Hewish Okoye Pulsar'. Interestingly, the very same supernova that created the Hewish Okoye Pulsar was also observed by astronomers, a thousand years earlier. The great supernova of 1054, the brightest in recorded history, was observed by Chinese astronomers and, as shown by a famous drawing on an overhanging cliff edge, by the peoples of Chaco Canyon in the south-western United States.

We haven't yet said how those neutrons manage to fend off gravity and prevent further collapse, but you can probably guess how it works. The neutrons (just like electrons) are slaves to the Pauli principle. They too can halt further collapse and so, just like white dwarves, neutron stars represent a possible end-point in the life of stars. Neutron stars are a detour as far as our story goes, but we can't leave them without remarking that these are very special objects in our wonderful Universe: they are stars the size of cities, so dense that a teaspoonful weighs as much as a mountain, held up by nothing more than the natural aversion to one another of spin-half particles.

There is only one option remaining for the most massive stars in the Universe – stars in which even the neutrons are moving close to light-speed. For such giants, disaster awaits, because the neutrons are no longer able to generate sufficient pressure to resist gravity. There is no known physical mechanism to stop a stellar core with a mass of greater than around three times the mass of our Sun falling in on itself, and the result is a black hole: a place where the laws of physics as we know them break down. Presumably Nature's laws don't cease to operate, but a proper understanding of the inner workings of a black hole requires a quantum theory of gravity, and no such theory exists today.

It is time to get back on message and to focus on our twin goals of proving the existence of white dwarf stars and calculating the Chandrasekhar mass. We know how to proceed: we must balance the electron pressure with gravity. This is not going to be a calculation we can do in our heads, so it will pay to make a plan of action. Here's the plan; it's quite lengthy because we want to clear up some background detail first and prepare the ground for the actual calculation.

Step 1: We need to determine what the pressure inside the star is due to those highly compressed electrons. You might be wondering why we are not worrying about the other stuff inside the star – what about the nuclei and the photons? Photons are not subject to the Pauli principle and, given enough time, they'll leave the star in any case. They have no hope of fighting gravity. As for the nuclei, the half-integer spin nuclei are subject to Pauli's rule but (as we shall see) their larger mass means they exert a smaller pressure than do the electrons and we can safely ignore their contribution to the balancing game. That simplifies matters hugely – the electron pressure is all we need, and that is where we should set our sights.

Step 2: After we've figured out the electron pressure, we'll need to do the balancing game. It might not be obvious how we should go about things. It's one thing to say 'gravity pulls in and the electrons push out' but it is quite another thing to put a number on it.

The pressure is going to vary inside the star; it will be larger in the centre and smaller at the surface. The fact that there is a pressure gradient is crucial. Imagine a cube of star matter sitting somewhere inside the star, as illustrated in Figure 12.2. Gravity will act to draw the cube towards the centre of the star and we want to know how the pressure from the electrons goes about countering it. The pressure in the electron gas exerts a force on each of the six faces of the cube, and the force is equal to the pressure at that face multiplied by the area of the face. That statement is precise; until now we have been using the word 'pressure' assuming that we all have sufficient intuitive understanding that a gas at high pressure

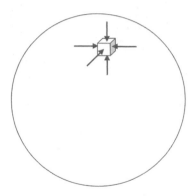

Figure 12.2. A small cube somewhere within the heart of a star. The arrows indicate the pressure exerted on the cube by the electrons within the star.

'pushes more' than a gas at low pressure. Anyone who has had to pump air into a flat car tyre knows that.

Since we are going to need to understand pressure properly, a brief diversion into more familiar territory is in order. Sticking with the tyre example, a physicist would say that a tyre is flat because the air pressure inside is insufficient to support the weight of the car without deforming the tyre: that's why we get to go to all the best parties. We can go ahead and calculate what the correct tyre pressure should be for a car with a mass of 1,500 kg if we want 5 centimetres of tyre to be in contact with the ground, as illustrated in Figure 12.3: it's chalk dust time again.

If the tyre is 20 cm wide and we want a 5 cm length of the tyre to be touching the road, then the area of tyre in contact with the ground will be $20 \times 5 = 100$ square centimetres. We don't know the requisite tyre pressure yet – this is what we want to calculate – so let's represent it by the symbol P. We need to know the downward force on the ground exerted by the air within the tyre. This is equal to the pressure multiplied by the area of tyre in contact with the floor, i.e. $P \times 100$ square centimetres. We should multiply that by four, because our car has four tyres: $P \times 400$ square centimetres. That is the total force exerted on the ground by the air within the tyres. Think of it like this: the air molecules inside the tyre are pound-

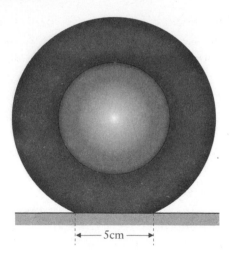

Figure 12.3. A tyre deforming slightly as it supports the weight of a car.

ing the ground (they are, to be pedantic, pounding the rubber in the tyre in contact with the ground, but that isn't important). The ground doesn't usually give way, in which case it pushes back with an equal but opposite force (so we did use Newton's third law after all). The car is being pushed up by the ground and pulled down by gravity and since it doesn't sink into the ground or leap into the air, we know that these two forces must balance each other. We can therefore equate the $P \times 400$ square centimetres of force pushing up with the downward force of gravity. That force is just the weight of the car and we know how to work that out using Newton's second law, $F = ma$, where a is the acceleration due to gravity at the Earth's surface, which is $9.81 \, \text{m/s}^2$. So the weight is $1{,}500 \, \text{kg} \times 9.8 \, \text{m/s}^2 = 14{,}700$ Newtons (1 Newton is equal to $1 \, \text{kg} \, \text{m/s}^2$ and it is roughly the weight of an apple). Equating the two forces implies that

$$P \times 400 \, \text{cm}^2 = 14{,}700 \, \text{N}$$

This is an easy equation to solve: $P = (14{,}700/400) \, \text{N/cm}^2 = 36.75 \, \text{N/cm}^2$. A pressure of 36.75 Newtons per square centimetre is probably not a very familiar way of stating a tyre pressure, but we can convert it into the more familiar 'bar'. 1 bar is standard air pressure, and is equal to $101,000$ Newtons per square metre. There are $10,000$ square centimetres in a square metre, so $101,000$ Newtons per square metre is equivalent to 10.1 Newtons per square centimetre. Our desired tyre pressure is therefore $36.75/10.1 = 3.6$ bar (or 52 psi – you can work that one out for yourself). We can also use our equation to deduce that, if the tyre pressure decreases by 50% to 1.8 bar, then we'll double the area of tyre in contact with the ground, which makes for a flatter tyre. After that refresher course on pressure we are ready to return to the little cube of star matter illustrated in Figure 12.2.

If the bottom face of the cube is closer to the centre of the star then the pressure on it should be a little bit bigger than the pressure pressing on the top face. That pressure difference gives rise to a force on the cube that wants to push the cube away from the centre of the star ('up' in the figure) and that is just what we want, because the cube will, at the same time, be pulled towards the centre of the star by gravity ('down' in the figure). If we could work out how to balance those two forces then we'd have developed some understanding of the star. But that is easier said than done because, although step 1 will allow us to work out how much the cube is pushed out by the electron pressure, we still have to figure out by how much gravity pulls in the opposite direction. By the way, we do not need to worry about the pressure pushing against the sides of our cube because the sides are equidistant from the centre of the star, so the pressure on the left side will balance the pressure on the right side and that ensures the cube does not move to the left or right.

To work out the force of gravity on the cube we need to make use of Newton's law of gravity, which tells us that every single piece of matter within the star pulls on our little cube by an amount that decreases in strength the farther the piece is from our cube. So more distant pieces pull less than closer ones. To deal with the fact that the gravitational pull on our cube is different for different pieces of star matter, depending on their distance away, looks like a tricky problem but we can see how to do it, in principle at least – we should chop the star into lots of pieces and then work out the force on the cube for each and every such piece. Fortunately, we do not need to imagine chopping the star up because we can exploit a very beautiful result. Gauss' law (named after the legendary German mathematician Carl Friedrich Gauss) informs us that: (a) we can totally ignore the gravity from all the pieces sitting further out from the centre of the star than our little cube; (b) the net gravitational effect of all of the pieces that sit closer to the centre is exactly as if all of those pieces were squashed together at the exact centre of the star. Using Gauss' law in conjunction with Newton's law of gravity we can say that the cube experiences a force that pulls it towards the centre of the star and that force is equal to

$$G\frac{M_{in}M_{cube}}{r^2}$$

where M_{in} is the mass of the star lying within a sphere whose radius reaches only as far out as the cube, M_{cube} is the mass of the cube and r is the distance of the cube from the star's centre (and G is Newton's constant). For example, if the cube sits on the surface of the star then M_{in} is the total mass of the star. For all other locations, M_{in} is smaller than that.

We're now making progress because to balance the forces on the cube (which we remind you means that the cube doesn't move and that means the star is not going to explode or collapse[2]) we require that

$$(P_{bottom} - P_{top})A = G\frac{M_{in}M_{cube}}{r^2} \tag{1}$$

where P_{bottom} and P_{top} are the pressures of the electron gas at the upper and lower faces of the cube and A is the area of each side of the cube (remember, the force exerted by a pressure is equal to the pressure multiplied by the area). We have labelled this equation '(1)' because it is very important and we will want to refer back to it.

---

2. We can generalize to the entire star because we are not being specific about where the cube actually is. If we can show that a cube located anywhere in the star does not move then that means all such cubes don't move and the star is stable.

Step 3: Make a cup of tea and feel pleased with ourselves because, after carrying out step 1, we will have figured out the pressures, P_{bottom} and P_{top}, and step 2 has made precise how to balance the forces. The real work is yet to come, though, because we still have to actually carry out step 1 and determine the pressure difference appearing on the left-hand side of equation (1). That is our next task.

Imagine a star packed with electrons and other stuff. How are the electrons scattered about? Let's focus our attention on a 'typical' electron. We know that electrons obey the Pauli Exclusion Principle, which means that no two electrons are likely to be found in the same region of space. What does that mean for the sea of electrons that we've been referring to as the 'electron gas' in our star? Because the electrons are necessarily separated from each other, we can suppose that each electron sits all alone inside a tiny imaginary cube within the star. Actually, that's not quite right because we know that electrons come in two types – 'spin up' and 'spin down' – and the Pauli principle only forbids identical particles from getting too close, which means we can fit two electrons inside a cube. This should be contrasted with the situation that would arise if the electrons did not obey the Pauli principle. In that case the electrons would not be localized two-at-a-time inside 'virtual containers'. Rather they could spread out and enjoy a much greater living space. In fact, if we were to ignore the various ways that the electrons can interact with each other and with the other particles in the star, there would be no limit to their living room.

We know what happens when we confine a quantum particle: it hops about according to Heisenberg's Uncertainty Principle, and the more it is confined the more it hops. That means that, as our would-be white dwarf collapses, so the electrons get increasingly confined and that makes them increasingly agitated. It is the pressure exerted by their agitation that will halt the gravitational collapse.

We can do better than words, because we can use Heisenberg's Uncertainty Principle to determine the typical momentum of an electron. In particular, if we confine the electron to a region of size Δx then it will hop around with a typical momentum $p \sim h/\Delta x$. Actually, in Chapter 4 we argued that this is more like an upper limit on the momentum and that the typical momentum is somewhere between zero and this value; that piece of information is worth remembering for later. Knowing the momentum allows us, immediately, to learn two things. Firstly, if the electrons didn't obey Pauli then they would not be confined to a region of size Δx but rather to some much larger size. That in turn would result in much less jiggling, and less jiggling means less pressure. So it is clear how the Pauli principle is entering the game; it is putting the squeeze on the electrons so that, via Heisenberg, they get a supercharged jiggle. In a moment we'll convert this idea of a supercharged jiggle into a formula for the pressure, but first we should mention the second thing we can learn. Because the momentum $p = mv$, the speed of the jiggle also depends inversely on the mass, so the electrons are jumping around much more vigorously than the heavier nuclei that also make up the star, and that is why the pressure exerted by the nuclei is unimportant. So how do we go from knowing the momentum of an electron to computing the pressure a gas of similar electrons exerts?

What we need to do first is to work out how big the little chunks containing the pairs of electrons must be. Our little chunks have volume $(\Delta x)^3$, and because we have to fit all the electrons inside the star, we can express this in terms of the number of electrons within the star (N) divided by the volume of the star (V). We'll need precisely $N/2$ containers to accommodate all of the electrons because we are allowed two electrons inside each container. This means that each container will occupy a volume of V divided by $N/2$, which is equal to $2(V/N)$. We'll need the quantity N/V (the number of electrons per unit volume inside the star) quite a lot in what follows, so we'll give it its own symbol n. We can now write down what the volume of the containers must be in order to contain all the electrons in the star, i.e. $(\Delta x)^3 = 2/n$. Taking the cube root of the right hand side allows us to conclude that

$$\Delta x = \sqrt[3]{2/n} = (2/n)^{1/3}$$

We can now plug this into our expression from the Uncertainty

Principle to get the typical momentum of the electrons due to their quantum jiggling:

$$p \sim h(n/2)^{1/3} \qquad (2)$$

where the $\sim$ sign means 'something like'. Clearly this is a bit vague because the electrons will not all be jiggling in exactly the same way: some will move faster than the typical value and some will move more slowly. The Heisenberg Uncertainty Principle isn't capable of telling us exactly how many electrons move at this speed and how many at that. Rather, it provides a more 'broad brush' statement and says if you squeeze an electron down then it will jiggle with a momentum something like $h/\Delta x$. We are going to take that typical momentum and assume it's the same for all the electrons. In the process, we will lose a little precision in our calculation but gain a great deal of simplicity as a result, and we are certainly thinking about the physics in the right way.[3]

We now know the speed of the electrons and that is enough information to work out how much pressure they exert on the tiny cube. To see that, imagine a fleet of electrons all heading in the same direction at the same speed (v) towards a flat mirror. They hit the mirror and bounce back, again travelling at the same speed but in the opposite direction. Let us compute the force exerted by the electrons on the mirror. After that we can attempt the more realistic calculation, where the electrons are not all travelling in the same direction. This methodology is very common in physics – first think about a simpler version of the problem you want to solve. That way you get to learn about the physics without biting off more than you can chew and gain confidence before tackling the harder problem. Imagine that the electron fleet consists of n particles per cubic metre and that, for the sake of argument, it has a circular cross-section of area 1 square metre – as illustrated in Figure 12.4. In one second nv electrons will hit the mirror (if v is measured in metres

---

3. It is of course possible to compute more precisely how the electrons move around but at the price of introducing more mathematics.

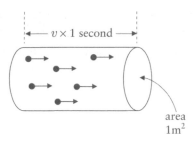

Figure 12.4. A fleet of electrons (the little dots) all heading in the same direction. All of the electrons in a tube this size will smash into the mirror every second.

per second). We know that because all of the electrons stretching from the mirror to a distance $v \times 1$ second away will smash into the mirror every second, i.e. all of the electrons in the tube drawn in the figure. Since a cylinder has a volume equal to its cross-sectional area multiplied by its length, the tube has a volume of v cubic metres and because there are n electrons per cubic metre in the fleet it follows that nv electrons hit the mirror every second.

When each electron bounces off the mirror it gets its momentum reversed, which means that each electron changes its momentum by an amount equal to $2mv$. Now, just as it takes a force to halt a moving bus and send it travelling back in reverse, so it takes a force to reverse the momentum of an electron. This is Isaac Newton once again. In Chapter 1 we wrote his second law as $F = ma$, but this is a special case of a more general statement, which states that the force is equal to the rate at which momentum changes.[4] So the whole fleet of electrons will impart a net force on the mirror $F = 2mv \times (nv)$, because this is the net change in momentum of the electrons every second. Due to the fact that the electron beam has an area of 1 square metre, this is also equal to the pressure exerted by the electron fleet on the mirror.

It is only a short step to go from a fleet of electrons to a gas of electrons. Rather than all the electrons ploughing along in the same

4. Newton's second law can be written as $F = dp/dt$. For constant mass this can be written in the more familiar form: $F = mdv/dt = ma$.

direction, we have to take into account that some travel up, some down, some to the left and so on. The net effect is to reduce the pressure in any one direction by a factor of 6 (think of the six faces on a cube) to $(2mv) \times (nv)/6 = nmv^2/3$. We can replace v in this equation by our Heisenberg-informed estimate of the typical speeds at which the electrons are zipping about (i.e. the previous equation (2)) to get the final result for the pressure exerted by the electrons in a white dwarf star:[5]

$$P = \frac{1}{3} nm \frac{h^2}{m^2} \left(\frac{n}{2}\right)^{2/3} = \frac{1}{3} \left(\frac{1}{2}\right)^{2/3} \frac{h^2}{m} n^{5/3}$$

If you recall, we said that this was only an estimate. The full result, using a lot more mathematics, is

$$P = \frac{1}{40} \left(\frac{3}{\pi}\right)^{2/3} \frac{h^2}{m} n^{5/3} \qquad (3)$$

This is a nice result. It tells us that the pressure at some place in the star varies in proportion to the number of electrons per unit volume at that place raised to the power of $5/3$. You should not be concerned that we did not get the constant of proportionality correct in our approximate treatment – the fact that we got everything else right is what matters. In fact, we did already say that our estimate of the momentum of the electrons is probably a little too big and this explains why our estimate of the pressure is bigger than the true value.

Knowing the pressure in terms of the density of electrons is a good start but it will suit our purposes better to express it in terms of the actual mass density in the star. We can do this under the very safe assumption that the vast majority of the star's mass comes from the nuclei and not the electrons (a single proton has a mass nearly 2,000 times greater than that of an electron). We also know that the number of electrons must be equal to the number of protons in the star because the star is electrically neutral. To get the mass density

---

5. Here we have combined the exponents according to the general rule $x^a x^b = x^{a+b}$.

we need to know how many protons and neutrons there are per cubic metre within the star and we should not forget the neutrons because they are a by-product of the fusion process. For lighter white dwarfs, the core will be predominantly helium-4, the end product of hydrogen fusion, and this means that there will be equal numbers of protons and neutrons. Now for a little notation. The atomic mass number, A, is conventionally used to count the number of protons + neutrons inside a nucleus and $A = 4$ for helium-4. The number of protons in a nucleus is given the symbol Z and for helium $Z = 2$. We can now write down a relationship between the electron density, n, and the mass density, ρ:

$$n = Z\rho/(m_pA)$$

and we've assumed that the mass of the proton, m_p, is the same as the mass of the neutron, which is plenty good enough for our purposes. The quantity m_pA is the mass of each nucleus; ρ/m_pA is then the number of nuclei per unit volume, and Z times this is the number of protons per unit volume, which must be the same as the number of electrons – and that's what the equation says.

We can use this equation to replace n in equation (3), and because n is proportional to ρ the upshot is that the pressure varies in proportion to the density to the power of $^5\!/_3$. The salient physics we have just discovered is that

$$P = \kappa\rho^{5/3} \tag{4}$$

and we should not be worrying too much about the pure numbers that set the overall scale of the pressure, which is why we just bundled them all up in the symbol κ. It's worth noting that κ depends on the ratio of Z and A, and so will be different for different kinds of white dwarf star. Bundling some numbers together into one symbol helps us to 'see' what is important. In this case the symbols could distract us from the important point, which is the relationship between the pressure and the density in the star.

Before we move on, notice that the pressure from quantum jiggling doesn't depend upon the temperature of the star. It only cares about how much we squeeze the star. There will also be an additional contribution to the electron pressure that comes about simply because the electrons are whizzing around 'normally' due to their temperature, and the hotter the star, the more they whizz around. We have not bothered to talk about this source of pressure because time is short and, if we were to go ahead and calculate it, we would find that it is dwarfed by the much larger quantum pressure.

Finally, we are ready to feed our equation for the quantum pressure into the key equation (1), which is worth repeating here:

$$(P_{bottom} - P_{top})A = G\frac{M_{in}M_{cube}}{r^2} \tag{1}$$

But this is not as easy as it sounds because we need to know the difference in the pressures at the upper and lower faces of the cube. We could re-write equation (1) entirely in terms of the density within the star, which is itself something that varies from place to place inside the star (it must be otherwise there would be no pressure difference across the cube) and then we could try to solve the equation to determine how the density varies with distance from the star's centre. To do this is to solve a differential equation and we want to avoid that level of mathematics. Instead, we are going to be more resourceful and think harder (and calculate less) in order to exploit equation (1) to deduce a relationship between the mass and the radius of a white dwarf star.

Obviously the size of our little cube and its location within the star are completely arbitrary, and none of the conclusions we are going to draw about the star as a whole can depend upon the details of the cube. Let's start by doing something that might seem pointless. We are quite entitled to express the location and size of the cube in terms of the size of the star. If R is the radius of the star, then we can write the distance of the cube from the centre of the star as $r = aR$, where a is simply a dimensionless number between 0 and 1. By dimensionless, we mean that it is a pure number and carries no units. If $a = 1$, the cube is at the surface of the star and if $a = \frac{1}{2}$ it is halfway out from the centre. Similarly, we can write the size of the cube in terms of the radius of the star. If L is the length of a side of the cube, then we can write $L = bR$ where, again, b is a pure number, which will be very small if we want the cube to be small relative to the star. There is absolutely nothing deep about this and, at this stage, it should seem so obvious as to appear pointless. The only noteworthy point is that R is the natural distance to use because there are no other distances relevant to a white dwarf star that could have provided any sensible alternatives.

Likewise, we can continue our strange obsession and express the density of the star at the position of the cube in terms of the average density of the star, i.e. we can write $\rho = f\bar{\rho}$ where f is, once again, a pure number and $\bar{\rho}$ is the average density of the star. As we have already pointed out, the density of the cube depends on its position inside the star – if it is closer to the centre, it will be more dense. Given that the average density $\bar{\rho}$ does not depend on the position of the cube, then f must do so, i.e. f depends on the distance r, which obviously means it depends on the product aR. Now, here is the key piece of information that underpins the rest of our calculation: f is a pure number and R is not a pure number (because it is measuring a distance). This fact implies that f can only depend upon a and not on R at all. This is a very important result, because it is telling us that the density profile of a white dwarf star is 'scale invariant'. This means the density varies with radius in the same way no matter what the radius of the star is. For example, the density at a point $3/4$ of the way out from the centre of the star will be the same fraction of the mean density in every white dwarf star, regardless of the star's size. There are two ways of appreciating this crucial result and we thought we'd present them both. One of us explained it thus: 'That's because any dimensionless function of r (which is what f is) can only be dimensionless if it is a function of a dimensionless variable, and the only dimensionless variable we have is $r/R = a$, because R is the only quantity which carries the dimensions of distance that we have at our disposal.'

The other author feels that the following is clearer: 'f can in general depend in a complicated way on r, the distance of the little cube from the centre of the star. But let's assume for the sake of this paragraph that it is directly proportional to it, i.e. $f \propto r$. In other words, $f = Br$, where B is a constant. Here, the key point is that we want f to be a pure number, whilst r is measured in (say) metres. That means that B must be measured in $1/$ metres, so that the units of distance cancel each other out. So what should we choose for B? We can't just choose something arbitrary, like '1 inverse metres', because this would be meaningless and has nothing to do with the star. Why not choose 1 inverse light years for example, and get a very different answer? The only distance we have to hand is R, the physical radius of the star, and so we are forced to use this to ensure that f will always be a pure number. This means that f depends only on r/R. You should be able to see that the same conclusion can be drawn if we started out by assuming that $f \propto r^2$ say.' Which is just what he said, only longer.

This means that we can express the mass of our little cube, of size L and volume L^3, sitting at a distance r from the centre of the star, as $M_{cube} = f(a)L^3\bar{\rho}$. We wrote $f(a)$ instead of just f in order to remind us that f really only depends upon our choice of $a = r/R$ and not on the large-scale properties of the star. The same argument can be used to say that we can write $M_{in} = g(a)M$ where $g(a)$ is again only a function of a. For example, the function $g(a)$ evaluated at $a = 1/2$ tells us the fraction of the star's mass lying in a sphere of half the radius of the star itself, and that is the same for all white dwarf stars, regardless of their radius because of the argument in the previous paragraph.[6] You might have noticed that we are steadily working our way through the various symbols which appear in equation (1), replacing them by dimensionless quantities $(a, b, f$ and $g)$ multiplied by quantities that depend only on the mass and radius of the star (the average density of the star is determined in terms of M and R because $\bar{\rho} = M/V$ and $V = 4\pi R^3/3$, the volume of a sphere). To complete the task, we just need to do the same

---

6. For those of a mathematical bent, show that $g(a) = 3\int_0^a x^2 f(x)\,dx$, i.e. that the function $g(a)$ is actually determined once we know the function $f(a)$.

for the pressure difference, which we can (by virtue of equation (4)) write as $P_{bottom} - P_{top} = h(a,b)\kappa\bar{\rho}^{5/3}$ where $h(a,b)$ is a dimensionless quantity. The fact that $h(a,b)$ depends upon both of a and b is because the pressure difference not only depends on where the cube is (represented by a) but also on how big it is (represented by b): bigger cubes will have a larger pressure difference. The key point is that, just like $f(a)$ and $g(a)$, $h(a,b)$ cannot depend upon the radius of the star.

We can make use of the expressions we just derived to rewrite equation (1):

$$(h\kappa\bar{\rho}^{5/3}) \times (b^2 R^2) = G\frac{(gM) \times (fb^3 R^3 \bar{\rho})}{a^2 R^2}$$

That looks like a mess and not much like we are within one page of hitting the jackpot. The key point is to notice that this is expressing a relationship between the mass of the star and its radius – a concrete relation between the two is within touching distance (or desperate grasping distance, depending on how well you handled the mathematics). After substituting in for the average density of the star (i.e. $\bar{\rho} = M/(4\pi R^3/3)$) this messy equation can be rearranged to read

$$RM^{1/3} = \kappa/(\lambda G) \tag{5}$$

where
$$\lambda = \left(\frac{4\pi}{3}\right)^{2/3}\frac{bfg}{ha^2}$$

Now λ only depends upon the dimensionless quantities a, b, f, g and h, which means that it does *not* depend upon the quantities that describe the star as a whole, M and R, and this means that it must take on the same value for all white dwarf stars.

If you are worrying what would happen if we were to change a and/or b (which means changing the locations and/or size of our little cube) then you have missed the power of this argument. Taken at face value, it certainly looks like changing a and b will change λ so that we will get a different answer for $RM^{1/3}$. But that is impossible, because we know that $RM^{1/3}$ is something that depends on the star and not on the specific properties of a little cube that we might or might not care to dream up. This means that any variation in a or b must be compensated for by corresponding changes in f, g and h.

Equation (5) says, quite specifically, that white dwarves can exist. It says that because we've been successfully able to balance the gravity–pressure equation (equation (1)). That is not a trivial thing – because it might have been possible that the equation could not be satisfied for any combination of M and R. Equation (5) also makes the prediction that the quantity $RM^{1/3}$ must be a constant. In other words, if we look up into the sky and measure the radius and the mass of white dwarves, we should find that the radius multiplied by the cube root of the mass will give the same number for every white dwarf. That is a bold prediction.

The argument that we just presented can be improved upon because it is possible to calculate exactly what the value of λ should be, but to do that we would need to solve a second-order differential equation in the density, and that is a mathematical bridge too far for this book. Remember, λ is a pure number: it simply 'is what it is' and we can, with a little higher-level maths, compute it. The fact that we did not actually work it out here should not detract at all from our achievements: we have proven that white dwarf stars can exist and we have managed to make a prediction relating their mass and radius. After calculating λ (which can be done on a home computer), and after substituting in the values for κ and G, the prediction is that

$$ RM^{1/3} = (3.5 \times 10^{17}\,\mathrm{kg^{1/3}m}) \times (Z/A)^{5/3} $$

which is equal to $1.1 \times 10^{17}\,\mathrm{kg^{1/3}m}$ for cores of pure helium, carbon or oxygen $(Z/A = 1/2)$. For iron cores, $Z/A = 26/56$ and the 1.1 reduces slightly to 1.0. We trawled the academic literature and collected together the data on the masses and radii of sixteen white dwarf stars sprinkled about the Milky Way, our galactic backyard. For each we computed the value of $RM^{1/3}$ and the result is that astronomical observations reveal $RM^{1/3} \approx 0.9 \times 10^{17}\,\mathrm{kg^{1/3}m}$. The agreement between the observations and theory is thrilling – we have succeeded in using the Pauli Exclusion Principle, the Heisenberg Uncertainty Principle and Newton's law of gravity to predict the mass–radius relationship of white dwarf stars.

There is, of course, some uncertainty on these numbers (the theory value of 1.0 or 1.1 and the observational number equal to 0.9). A proper scientific analysis would now start talking about just how likely it is that the theory and experiment are in agreement, but for our purposes that level of analysis is unnecessary because the agreement is already staggeringly good. It is quite fantastic that we have managed to figure all this out to an accuracy of something like 10%, and is compelling evidence that we have a decent understanding of stars and of quantum mechanics.

Professional physicists and astronomers would not leave things here. They would be keen to test the theoretical understanding in as much detail as possible, and to do that means improving on the description we presented in this chapter. In particular, an improved analysis would take into account that the temperature of the star does play some role in its structure. Furthermore, the sea of electrons is swarming around in the presence of positively charged atomic nuclei and, in our calculation, we totally ignored the interactions between the electrons and the nuclei (and between electrons and electrons). We neglected these things because we claimed that they would produce fairly small corrections to our simpler treatment. That claim is supported by more detailed calculations and it is why our simple treatment agrees so well with the data.

We have obviously learnt an awful lot already: we have established that the electron pressure is capable of supporting a white dwarf star and we have managed to predict with some precision how the radius of the star changes if we add or remove mass from the star. Unlike 'ordinary' stars that are eagerly burning fuel, notice that white dwarf stars have the feature that adding mass to a star makes it smaller. This happens because the extra stuff we add goes into increasing the star's gravity, and that makes it contract. Taken at face value the relationship expressed in equation (5) seems to imply that we would need to add an infinite amount of mass before the star shrinks to no size at all. But this isn't what happens. The important thing, as we mentioned at the beginning of the chapter, is that we eventually move into the regime where the electrons are so tightly packed that Einstein's Theory of Special Relativity becomes important because the speed of the electrons starts to approach the speed of light. The impact on our calculation is that we have to stop using Newton's laws of motion, and replace them with Einstein's laws. This, as we shall now see, makes all the difference.

What we're about to find is that as the star gets more massive, the pressure exerted by the electrons will no longer be proportional to the density raised to the power $\frac{5}{3}$; instead, the pressure increases less quickly with density. We will do the calculation in a moment, but straight away we can see that this could have catastrophic consequences for the star. It means that when we add mass, there will be the usual increase in gravity but a smaller increase in pressure. The star's fate hinges on just how much 'less quickly' the pressure varies with density when the electrons are moving fast. Clearly it is time to figure out what the pressure of a 'relativistic' electron gas is.

Fortunately, we do not need to wheel in the heavy machinery of Einstein's theory because the calculation of the pressure in a gas of electrons moving close to light speed follows almost exactly the same reasoning as that we just presented for a gas of 'slow-moving' electrons. The key difference is that we can no longer write that the momentum $p = mv$, because this is not correct any more. What is correct, though, is that the force exerted by the electrons is still equal to the rate of change of their momentum. Previously, we deduced that a fleet of electrons bouncing off a mirror exerts a pressure $P = 2mv \times (nv)$. For the relativistic case, we can write the same expression, but providing that we replace mv by the momentum, p. We are also assuming that the speed of the electrons is close to the speed of light, so we can replace v with c. Finally, we still have to divide by 6 to get the pressure in the star. This means that we can write that the pressure for the relativistic gas as $P = 2p \times nc/6 = pnc/3$.

Just as before, we can now go ahead and use Heisenberg's Uncertainty Principle to say that the typical momentum of the confined electrons is $h(n/2)^{1/3}$ and so

$$P = \frac{1}{3}nch\left(\frac{n}{2}\right)^{1/3} \propto n^{4/3}$$

Again we can compare this to the exact answer, which is

$$P = \frac{1}{16}\left(\frac{3}{\pi}\right)^{1/3} hcn^{4/3}$$

Finally, we can follow the same methodology as before to express the pressure in terms of the mass density within the star and derive the alternative to equation (4):

$$P = \kappa' \rho^{4/3}$$

where $\kappa' \propto hc \times (Z/(Am_p))^{4/3}$. As promised, the pressure increases less quickly as the density increases than it does for the non-relativistic case. Specifically, it increases with a power of $\frac{4}{3}$ rather than $\frac{5}{3}$. The reason for this slower variation can be traced back to the fact that the electrons cannot travel faster than the speed of light. This means that the 'flux' factor, nv, which we used to compute the pressure saturates at nc and the gas is not capable of delivering the electrons to the mirror (or face of the cube) at a sufficient rate to maintain the $\rho^{5/3}$ behaviour.

We can now explore the implications of this change because we can go through the same argument as in the non-relativistic case to derive the counterpart to equation (5):

$$\kappa' M^{4/3} \propto GM^2$$

This is a very important result because, unlike equation (5), it does not have any dependence upon the radius of the star. The equation is telling us that this kind of star, packed with light-speed electrons, can only have a very specific value of its mass. Substituting in for κ' from the previous paragraph gives us the prediction that

$$M \propto \left(\frac{hc}{G}\right)^{3/2} \left(\frac{Z}{Am_p}\right)^2$$

This is exactly the result we advertised right at the start of this chapter for the maximum mass that a white dwarf star can possibly have. We are very close to reproducing Chandrasekhar's result. All that remains to understand is why this special value is the maximum possible mass.

We have learnt that for white dwarf stars that are not too massive, the radius is not too small and the electrons are not too squashed. They therefore do not quantum jiggle to excess and their speeds are small compared to the speed of light. For these stars, we have seen that they are stable with a mass–radius relationship of the form $RM^{1/3}$ = constant. Now imagine adding more mass to the star. The mass–radius relation informs us that the star shrinks and, as a result, the electrons are more compressed and that means they jiggle faster. Add yet more mass and the star shrinks some more. Adding mass therefore increases the speed of the electrons until, eventually, they are travelling at speeds comparable with the speed of light. At the same time, the pressure will slowly change from $P \propto \bar{\rho}^{5/3}$ to $P \propto \bar{\rho}^{4/3}$ and in the latter case, the star is only stable at one particular value of the mass. If the mass is increased beyond this specific value then the right-hand side of $\kappa' M^{4/3} \propto GM^2$ becomes larger than the left-hand side and the equation is unbalanced. This means that the electron pressure (which resides on the left-hand side of the equation) is insufficient to balance the inward pull of gravity (which resides on the right-hand side) and the star must necessarily collapse.

If we were more careful with our treatment of the electron momentum and had taken the trouble to wheel in the advanced mathematics to compute the missing numbers (again a minor task for a personal computer), we could make a precise prediction for the maximum mass of a white dwarf star. It is

$$M = 0.2 \left(\frac{hc}{G}\right)^{3/2} \left(\frac{Z}{Am_p}\right)^2 = 5.8 \left(\frac{Z}{A}\right)^2 M_\odot,$$

where we have re-expressed the bundle of physical constants in terms of the mass of our Sun ($M_\odot$). Notice, by the way, that all the extra hard work that we have not done simply returns the constant of proportionality, which has a value of 0.2. This equation delivers the sought-after Chandrasekhar limit: 1.4 solar masses for $Z/A = 1/2$.

This really is the end of our journey. The calculation in this chapter has been at a higher mathematical level than the rest of the book but it is, in our view, one of the most spectacular demonstrations of the sheer power of modern physics. To be sure, it is not a 'useful' thing, but it is surely one of the great triumphs of the human mind. We used relativity, quantum mechanics and careful mathematical reasoning to calculate correctly the maximum size of a blob of matter that can be supported against gravity by the Exclusion Principle. This means that the science is right; that quantum mechanics, no matter how strange it might seem, is a theory that describes the real world. And that is a good way to end.

Further Reading

We used many books in the preparation of this book, but some deserve special mention and are highly recommended.

For the history of quantum mechanics, the definitive sources are two superb books by Abraham Pais: *Inward Bound* and *Subtle Is the Lord* . . . Both are quite technical but they are unrivalled in historical detail.

Richard Feynman's book *QED: The Strange Theory of Light and Matter* is at a similar level to this book and is more focused, as the title suggests, on the theory of quantum electrodynamics. It is a joy to read, like most of Feynman's writings.

For those in search of more detail, the very best book on the fundamentals of quantum mechanics is, in our view, still Paul Dirac's book *The Principles of Quantum Mechanics*. A high level of mathematical ability is needed to tackle this one.

Online, we should like to recommend two lecture courses that are available on iTunes University: Leonard Susskind's 'Modern Physics: The Theoretical Minimum – Quantum Mechanics' and James Binney's more advanced 'Quantum Mechanics' from the University of Oxford. Both require a reasonable mathematical background.

Index

Index